Progress in Probability

Volume 17

Seminar on Stochastic Processes, 1988

E. Çınlar
K.L. Chung
R.K. Getoor
Editors

J. Glover
Managing Editor

1989

Birkhäuser
Boston · Basel · Berlin

E. Çınlar
Department of Civil Engineering and Operations Research
Princeton University
Princeton, NJ 08544
U.S.A.

K.L. Chung
Department of Mathematics
Stanford University
Stanford, CA 94305
U.S.A.

R.K. Getoor
Department of Mathematics
University of California, San Diego
La Jolla, CA 92093
U.S.A.

J. Glover (Managing Editor)
Department of Mathematics
University of Florida
Gainesville, FL 32611
U.S.A.

ISSN: 0892-063X

Library of Congress Cataloging-in-Publication Data
Seminar on Stochastic Processes (8th : 1988 : University of Florida)
Seminar on Stochastic Processes, 1988 / E. Çinlar, K.L. Chung, R.K. Getoor, editors ; J. Glover, managing editor.
p. cm.—(Progress in probability and statistics ; v. 17)
Proceedings of the Eighth Seminar on Stochastic Processes held at the University of Florida, Gainesville, March 3–5, 1988.
Includes bibliographies and index.

ISBN-13: 978-1-4612-8217-4 e-ISBN-13: 978-1-4612-3698-6
DOI: 10.1007/978-1-4612-3698-6

1. Stochastic processes—Congresses. I. Çinlar, E. (Erhan), 1941– . II. Chung, Kai Lai, 1917– . III. Getoor, R.K. (Ronald) Kay), 1929– . IV. Glover, J. (Joseph) V. Title. VI. Series.
QA274.A1S44 1988
519.2—dc19 88-39438
CIP

Printed on acid-free paper.

Softcover reprint of the hardcover 1st edition 1989

Prepared by the authors in camera-ready form.

9 8 7 6 5 4 3 2 1

FOREWORD

The 1988 Seminar on Stochastic Processes was held at the University of Florida, Gainesville, March 3 through March 5, 1988. It was the eighth seminar in a continuing series of meetings which provide opportunities for researchers to discuss current work in stochastic processes in an informal and enjoyable atmosphere. Previous seminars were held at Princeton University, Northwestern University, the University of Florida and the University of Virginia. The participants' enthusiasm and interest have created stimulating and successful seminars. We thank those participants who have permitted us to publish their research in this volume. This year's invited participants included B. Atkinson, J. Azema, D. Bakry, P. Baxendale, J. Brooks, G. Brosamler, K. Burdzy, E. Cinlar, R. Darling, N. Dinculeanu, E. Dynkin, S. Evans, N. Falkner, P. Fitzsimmons, R. Getoor, J. Glover, V. Goodman, P. Hsu, J.-F. Le Gall, M. Liao, P. March, P. McGill, J. Mitro, T. Mountford, C. Mueller, A. Mukherjea, V. Papanicolaou, E. Perkins, M. Pinsky, L. Pitt, A. O. Pittenger, Z. Pop-Stojanovic, M. Rao, J. Rosen, T. Salisbury, C. Shih, M. Taksar, J. Taylor, S. J. Taylor, E. Toby, R. Williams, Wu Rong, and Z. Zhao. The seminar was made possible through the generous support of the Department of Mathematics, the Center for Applied Mathematics, the Division of Sponsored Research and the College of Liberal Arts and Sciences of the University of Florida. We extend our thanks for local arrangements to our host, Zoran Pop-Stojanovic.

J. G.

Gainesville, 1988

TABLE OF CONTENTS

D. Bakry.....The Riesz Transform Associated with Second Order Differential Operators 1

J. K. Brooks & D. Neal.....The Optional Stochastic Integral 45

K. Burdzy, E. H. Toby, & R. J. Williams.....On Brownian Excursions in Lipschitz Domains II: Local Asymptotic Distributions 55

K. L. Chung.....Gauge Theorem for Unbounded Domains 87

K. L. Chung.....Reminiscences of Some of Paul Levy's Ideas in Brownian Motion and in Markov Chains 99

M. Cranston.....Conditional Brownian Motion, Whitney Squares, and the Conditional Gauge Theorem 109

S. N. Evans.....Local Field Gaussian Measures 121

P. J. Fitzsimmons & R. K. Getoor.....Some Formulas for the Energy Functional of a Markov Process 161

I. W. Herbst & Z. Zhao.....Note on the 3G Theorem (d=2) 183

H. R. Hughes & M. Liao.....The Independence of Hitting Times and Hitting Positions to Spheres for Drifted Brownian Motion 185

J.-F. Le Gall.....The Exact Hausdorff Measure of Brownian Multiple Points II. 193

P. March.....On a Stability Property of Harmonic Measures 199

Z. R. Pop-Stojanovic.....Behavior of Excessive Functions of Certain Diffusions Under the Action of the Transition Semigroup 213

M. Rao.....A Maximal Inequality 221

M. Rao.....Some Results for Functions of Kato Class in Domains of Infinite Measure 225

R. Wu.....Some Properties of Invariant Functions of Markov Processes 239

Z. Zhao.....Right Brownian Motion and Representation of Initial Problem 245

The RIESZ transforms associated with second order differential operators.

Dominique Bakry*

This is an expository paper about some recent works on the RIESZ transforms associated with a general symmetric, elliptic second order differential operator. Except in the third part, there are no new results and the proofs are often ommited or just sketched.

In the first chapter, we present the classical results on RIESZ transforms : on the circle, on the real line and on $\mathcal{R}^n$. We also present some results which are less classical but present some interest : for the spheres, for the ultraspherical operators and for the ORNSTEIN-UHLENBECK operator. There are many other results that we did not mention, for example on compact semisimple LIE groups, on CARTAN-HADAMARD manifolds, and for the HEISENBERG group. (The reader interested in these results is referred to the bibliography.)

In the second part, we introduce the basic concepts of differential geometry that we will use in the sequel. We try to be understandable by a reader who is unfamiliar with these notions, and we explicit all the concepts that we are going to work on. In particular, we introduce the notion of RICCI curvature associated with a second order differential operator, which extends the notion of RICCI curvature of a Riemannian manifold. We also introduce a whole family of "harmonic extensions" to the half space of functions defined on our space, and we show how to construct them with the heat semigroup.

In the third part, we deal with the L^p theory of RIESZ transforms. We first give comparison results (expressed in terms of boundedness of some LAPLACE transforms) and then we give a common proof for results in different contexts : when the RICCI curvature is bounded

* This work was done while the author was visiting the University of British Columbia

away from 0, when it is positive, and when it is just bounded from below.

The last part is devoted to the $\mathbf{H}^1$ theory. We introduce a notion of dimension associated to a second order differential operator, and we use this notion to give a general subharmonicity lemma. We then use this lemma to state a partial result in $\mathbf{H}^1$ theory when the dimension is finite and the curvature is bounded from below. When the RICCI curvature is positive, we give a complete norm equivalence analoguous to the classical one.

1.— Classical results.

a)— The HILBERT transform on the circle and the real line

On the unit cercle $\mathbf{T}$ or on the real line $\mathcal{R}$, the RIESZ transform is nothing but the classical HILBERT transform. Let us describe it on the circle, for example. This is the operator wich is defined on smooth functions by

$$\mathcal{H}f(e^{ix}) = \lim_{\varepsilon \to 0} \int_{|t|>\varepsilon} f(e^{i(x-t)}) \cot(\frac{t}{2})\, dt.$$

This operator appears when one studies the holomorphic functions on the disk $\{|z| < 1\}$. To be precise, consider a nice real function f on the circle , say in the $\mathcal{C}^\infty$ class. Let $\hat{f}$ be the harmonic prolongation of f to the disk, i.e.

$$\hat{f}(re^{ix}) = \int_{\mathbf{T}} f(e^{i(x-y)}) p(r, e^{iy}) dy$$

where $p(r, e^{iy})$ is the POISSON kernel $p(r, e^{iy}) = \dfrac{1-r^2}{1-2r\cos y + r^2}$.
Then, the function $\hat{f}$ is the real part of an harmonic function $H(z) = \hat{f} + i\hat{g}$. The function $\hat{g}$ is itself the harmonic prolongation of a real function g wich is defined on the circle: **this function g is the HILBERT transform of the function f.**

Another way to look at the HILBERT transform is to consider it's decomposition in FOURRIER series.

$$\text{If } f(e^{i\theta}) = \sum_{n\in\mathbf{Z}} a_n e^{in\theta}, \quad \text{with} \quad a_{-n} = \bar{a}_n,$$

$$\text{then, one has } \mathcal{H}f(e^{i\theta}) = \sum i \,\mathrm{sign}(n) a_n\, e^{in\theta},$$

where $\mathrm{sign}(n)$ is 1 if $n > 0$, is -1 if $n < 0$, and $\mathrm{sign}(0) = 0$.

In other words, consider the LAPLACE operator $\Delta = \dfrac{\partial^2}{\partial\theta^2}$ on the circle : in the space $\mathbf{L}^2$, this is a negative selfadjoint operator, whose spectral decomposition is given by the FOURRIER decomposition : we have

$$\Delta\Big(\sum_{n\in\mathbf{Z}} a_n e^{in\theta}\Big) = \sum_{n\in\mathbf{Z}} -n^2 a_n e^{in\theta}.$$

The operator $-\Delta$ has a symmetric positive square root $\mathbf{C}$ which is given by

$$\mathbf{C}\Big(\sum_{n\in\mathbf{Z}} a_n e^{in\theta}\Big) = \sum_{n\in\mathbf{Z}} |n| a_n e^{in\theta}.$$

The operator $\mathbf{C}^{-1}$ is then defined on functions with mean 0 by

$$\mathbf{C}^{-1}\Big(\sum_{n\neq 0} a_n e^{in\theta}\Big) = \sum_{n\neq 0} \frac{1}{|n|} a_n e^{in\theta}.$$

On this last formula, it is clear that the HILBERT transform stands for $-\dfrac{\partial}{\partial\theta}\Big(\mathbf{C}^{-1}\Big)$. On this expression, we can see that $\mathbf{C}$ is an operator of order 0.

The HILBERT transform has the following properties :

1. It is an isometry of the subspace $\mathbf{L}^2_0$ of $\mathbf{L}^2$ consisting of functions with mean 0. We have $\mathcal{H}^2 = -Id$.

2. (Theorem of M.Riesz.) For all p in $(1, \infty)$ $\mathcal{H}$ is a bounded operator on $\mathbf{L}^p_0$, the space of mean 0 functions in $\mathbf{L}^p$.

3. $\mathcal{H}$ is not bounded in $\mathbf{L}^1_0$ or in $\mathbf{L}^\infty_0$; in order to understand what happens in these cases, let us introduce a few notions.

We shall say that a function f is in $\mathbf{H}^1$ if it is in $\mathbf{L}_0^1$ together with it's HILBERT transform $\mathcal{H}f$. We shall set

$$\|f\|_{H^1} = \|f\|_1 + \|\mathcal{H}f\|_1.$$

Consider then the POISSON integral $\hat{f}$ of f. This is an harmonic function in the disk. If B_t is a brownian motion in $\mathcal{R}^2$ starting from 0, and if we denote by T is the first exit time from the disk, $M_t = \hat{f}(B_{t\wedge T})$ is a local martingale. The fundamental property of the HILBERT transform is the following :

Theorem .— Let f be a function in $\mathbf{L}_0^1$. Then, it is in the space $\mathbf{H}^1$ if and only if the associated martingale M_t is in the probabilistic $\mathbf{H}^1$ space, which means that $\sup_t |M_t|$ is in $\mathbf{L}^1$. Moreover, the norms $\|f\|_{\mathbf{H}^1}$ and $\|\sup_t |M_t|\|_1$ are equivalent.

The theorem of M.RIESZ is a (not too hard) consequence of the fact that, on the circle, the operators $\dfrac{\partial}{\partial\theta}$ and Δ commute. In order to understand how the $\mathbf{H}^1$ theorem works, let us sketch the proof of the inequality $\|\sup_t |M_t|\|_1 \leq C\,\|f\|_{\mathbf{H}^1}$, which is the hardest one.

If we denote by g the HILBERT transform of f, and by $\hat{f}$ and $\hat{g}$ the harmonic prolongations of f and g, then we know that the function $h = \hat{f} + i\hat{g}$ is holomorphic in the disk. We can then use a well known and useful property of holomorphic functions : *the logarithm of the modulus of an holomorphic function is an harmonic function.*

Therefore, for each real $q > 0$, the function $|h|^q$ is sub-harmonic in the disk, which impies that the process $Z_t = |h|^q(B_{t\wedge T})$ is a sub-martingale. Hence, choosing any q in $(0, 1)$, we may use DOOB's inequality to assert that

$$E[\sup_{t\leq T} |Z_t|^{1/q}] \leq C(q)E[|Z_T|^{1/q}].$$

But $|Z_t|^{1/q}$ majorizes $|M_t|$, where M_t is the martingale associated to f, while $|Z_T|^{1/q} = [f(B_T)^2 + g(B_T)^2]^{1/2}$. It remains to remark that B_T is uniformely distributed on the circle to see that $E[|Z_T|^{1/q}]$ is dominated by $\|f\|_{\mathbf{H}^1}$.

On the real line, one can rephrase all the results stated on the circle. There, the HILBERT transform is given by the singular kernel $1/\pi(x-y)$, which means that one has

$$\mathcal{H}f(x) = \lim_{\varepsilon \to 0} \int_{|x-y|>\varepsilon} \frac{f(y)}{\pi(x-y)}\, dy\, ;$$

By FOURRIER transform, one gets $\widetilde{\mathcal{H}f}(x) = -i(sgx)\tilde{f}(x)$ where $\tilde{f}$ denotes the FOURRIER transform of a function f. Once again, the HILBERT transform is related to harmonic prolongations, this time to the upper half-plane : $\mathcal{H}f$ is the boundary value of the harmonic conjuguate of the harmonic prolongation of f. It is still equal to $\frac{\partial}{\partial x}\circ(\frac{-\partial^2}{\partial x^2})^{-1/2}$; it shares the same properties that the HILBERT transform on the circle, the main difference being that, since the measure of the line is infinite, one does not need to restrict himself to the functions with mean 0. For the analoguous of the results in $\mathbf{L}^1$, we have to replace the Brownian motion starting from 0 in the disk by a Brownian motion starting from a point (t, X_0) in the upper half-plane, with $t > 0$ and X_0 uniformely distributed, and we stop it when it hits the real axis. Then we let t tends to infinity : we get the "white noise of the universe" of D.GUNDY.

b)— The RIESZ transforms on $\mathcal{R}^n$

The classical RIESZ transforms on $\mathcal{R}^n$ are the analoguous of the HILBERT transform on the real line or on the circle ; there are n of them, defined by

$$\mathbf{R}^i(f) = \frac{\partial}{\partial x^i}\circ(-\Delta)^{-1/2}(f),$$

where, as before, the square root of Δ has to be understood in the sense of $\mathbf{L}^2(\mathcal{R}^n)$. In terms of FOURRIER transforms, one has

$$\widetilde{\mathbf{R}^j(f)}(\xi) = i\frac{\xi^j}{|\xi|}\tilde{f}(\xi).$$

They also are defined by mean of singular integral kernels, namely $\mathbf{R}^i$ is the convolution with the function $\Omega^i(x)/|x|^n$, where $\Omega^i(x) = c_n\frac{x^i}{|x|}$, with a normalizing constant c_n whose value is unimportant for us.

This is the basic example of CALDERON-ZYGMUND operator, and in fact the RIESZ transforms are the fundamental bricks with which one builts all the singular integral operators in $\mathcal{R}^n$. (*cf* [S], for example.)

The interpretation of the RIESZ transforms in terms of holomorphic functions is still valid, once we have defined what will play that role. In an open set of $\mathcal{R}^n$, this notion is replaced by a system $(f_1, \cdots, f_n)$ of $\mathcal{C}^\infty$ functions satisfying the CAUCHY-RIEMANN equations :

$$(1) \qquad \begin{cases} \dfrac{\partial f^i}{\partial x^j} = \dfrac{\partial f^j}{\partial x^i} \\ \displaystyle\sum_i \frac{\partial f^i}{\partial x^i} = 0 \end{cases}$$

Now, let us consider the POISSON kernel on $\mathcal{R}^n$, which associates to each bounded function f on $\mathcal{R}^n$ its harmonic extension $\hat{f}$ on $\mathcal{R}^n \times \mathcal{R}_+$: we have

$$\hat{f}(x) = \int f(y) p(x,y)\, dy \quad \text{where} \quad p(x,y) = \frac{c_n t}{(t^2 + |x_0 - y|^2)^{(n+1)/2}}$$

at a point $x = (x_0, t) \in \mathcal{R}^n \times \mathcal{R}_+$. (Here, $c_n = \Gamma(\frac{n+1}{2})/\pi^{\frac{n+1}{2}}$, such that $\int p(x,y)\, dy = 1$.)

The connection between the RIESZ transforms and the POISSON kernel is the following : let $\widehat{f^0}$ be the harmonic prolongation of f and let $\widehat{f^i}$ be the harmonic prolongation of it's RIESZ transforms. Then, the system $(\widehat{f^0}, \widehat{f^1}, \cdots, \widehat{f^n})$ satisfy the CAUCHY-RIEMANN equations in $\mathcal{R}^n \times \mathcal{R}_+$. (Here, the variable x_0 stands for the extra variable in $\mathcal{R}_+$.)

As on the circle, the RIESZ transforms satisfy the following properties :

1. They are bounded operators on the space $\mathbf{L}^2$. We have

$$\sum_i (\mathbf{R}^i)^2 = -\mathbf{I}.$$

2. For all p in $(1, \infty)$ $\mathbf{R}^i$ is a bounded operator on $\mathbf{L}^p(\mathcal{R}^n)$. There is a norm equivalence

$$c_p \|f\|_p \leq \| \Big(\sum_i (\mathbf{R}^i f)^2 \Big)^{1/2} \|_p \leq C_p \|f\|_p \,,$$

with constants c_p and C_p independant of the dimension. (One can find a rather elementary probabilistic proof of that in the notes of MEYER [M1]). Once again, it relies heavily on the fact that the LAPLACE operator Δ commutes with the partial derivatives $\dfrac{\partial}{\partial x^i}$, which property will not be preserved in further extensions.

3. As in the case of the circle or of the real line, $\mathbf{R}^i$ is not bounded in $\mathbf{L}^1(\mathcal{R}^n)$ nor in $\mathbf{L}^\infty(\mathcal{R}^n)$; as before, one has to introduce a space $\mathbf{H}^1$ to deal with functions which are in $\mathbf{L}^1(\mathcal{R}^n)$ together with their RIESZ transforms. We set

$$\|f\|_{H^1} = \|\left[f^2 + \sum_i (\mathbf{R}^i f)^2\right]^{1/2}\|_1.$$

Among other characterization of the space $\mathbf{H}^1$, we have as before a probabilistic one : for a function f in $\mathbf{L}^1(\mathcal{R}^n)$, let $\hat{f}$ be it's harmonic prolongation to the half- space $\mathcal{R}^n \times \mathcal{R}_+$. If B_t is a Brownian motion in $\mathcal{R}^{n+1}$, stopped at the first exit time of $\mathcal{R}^n \times \mathcal{R}_+$, then $M_t = \hat{f}(B_t)$ is a local martingale. Let now B_0 be uniformely distributed on the hyperplane $\{t = a > 0\}$, where the variable t stands for the extra coordinate in $\mathcal{R}_+$. If we denote by $E^a(\cdot)$ the expectation under this initial law (which is not really an expectation under a probability measure since our initial law is infinite), we write $\|f\|_{H^1_a} = E^a[\sup_t |M_t|]$. Then, we have an norm equivalence

$$c\|f\|_{H^1} \leq \lim_{a\to\infty} \|f\|_{H^1_a} \leq C(n)\|f\|_{H^1}.$$

In this previous inequality, if one menber is infinite, such is the other one. The constant c is universal, and the constant $C(n)$ depends only on the dimension n (and tends to infinity with n).

Remark.—

One may reformulate the results in $\mathbf{L}^p(\mathcal{R}^n)$ and in $\mathbf{H}^1$ in a more compact way if we introduce the vector-valued operator $\vec{\mathbf{R}} = \nabla(-\Delta)^{-1/2}$ whose components are the operators $\mathbf{R}^i$. The result in $\mathbf{L}^p(\mathcal{R}^n)$ tells us that the function $|\vec{\mathbf{R}}f|$ has a norm in $\mathbf{L}^p(\mathcal{R}^n)$ which is equivalent to the norm of f, and the result in $\mathbf{H}^1$ compares the norm of $(f^2 + |\vec{\mathbf{R}}f|^2)^{1/2}$ in $\mathbf{L}^1(\mathcal{R}^n)$ to the probalistic $\mathbf{H}^1$ norm of the martingale constructed with the Brownian motion arriving from infinity.

In the case of $\mathcal{R}^n$, the $\mathbf{H}^1$ norm inequality does no longer rely on the properties of holomorphic functions, but there is an analogous property of the system satisfying the CAUCHY-RIEMANN equations which will play the same role for us, namely **the sub-harmonicity property of STEIN and WEISS** [SW]. This property asserts that, if a system $(f^1, \cdots, f^n)$ satisfy the CAUCHY-RIEMANN equations in an open set of $\mathcal{R}^n$, then, one has

$$(2) \qquad \Delta\Big[\sum_i (f^i)^2\Big]^{p/2} \geq 0, \text{ as soon as } \quad p \geq \frac{n-2}{n-1}.$$

In fact, one may reformulate this property in a way which is more accurate for further extensions.

First, observe that, locally, a system satisfying the CAUCHY-RIEMANN equations is nothing but the gradient of an harmonic function : the first equation in (1) tells us that the vector $(f^1, \cdots, f^n)$ has curl 0, so that it is locally the gradient of a function F, and the second equation tells us that this function F is harmonic.

Therefore, the sub-harmonicity lemma of STEIN and WEISS can be restated in the following way :

Lemma .—If F is an harmonic function in an open set Ω of $\mathcal{R}^n$, then $\Delta\big(|\nabla F|^p\big) \geq 0$, as soon as $p \geq (n-1)/(n-2)$.

c)—The RIESZ transforms on the spheres.

If we want to replace $\mathcal{R}^n$ by an n-dimensionnal sphere $\mathbf{S}^n$, as we did in the case of the real line, the first difficulty is that there are many possible extensions of the notion of RIESZ transforms. Let us show at least two natural definitions :

The first method is to consider the sphere as a Riemannian manifold and to replace the LAPLACE operator by the LAPLACE-BELTRAMI operator $\Delta_{\mathbf{S}^n}$ of the sphere. This is a negative self-adjoint operator on the space $\mathbf{L}^2(\mathbf{S}^n)$; therefore, we may consider the vector valued operator $\vec{\mathbf{R}} = \nabla(-\Delta_{\mathbf{S}^n})^{-1/2}$.

In order to give to this a reasonable meaning, we have to restrict ourselves to the functions on the sphere with mean 0 : in fact, the operator Δ has a discrete spectrum and the eigenspace associated to the eigenvalue 0 is the space of constant functions(*), whose orthogonal complement is the space of mean 0 functions. On this space, the operator $(-\Delta)^{-1/2}$ is a bounded operator, which maps $\mathcal{C}^\infty$ functions onto $\mathcal{C}^\infty$ functions. Hence, $\vec{\mathbf{R}}$ is a perfectly well defined operator on the space of smooth functions with mean 0. It is this operator we considered in [B1], for example, for which we prooved the analoguous of the results stated above in $\mathcal{R}^n$. Let us call for a moment this RIESZ transform the **geometric** one.

Another way to generalize the notion of RIESZ transform to the sphere is to consider the sphere $\mathbf{S}^n$ as the boundary of the ball of radius 1 in $\mathcal{R}^{n+1}$ and to adopt the point of vue of the CAUCHY-RIEMANN equations. From this point of vue, given a continuous function f on the sphere, one has to consider it's harmonic prolongation $\hat{f}$ inside the ball. Then, we look for n harmonic functions inside the ball such that, together with the function $\hat{f}$, they satisfy the CAUCHY-RIEMANN equations. This means that we are looking for an harmonic function $\hat{F}$ in the ball such are our $(n+1)$ functions are the different components of it's gradient. But, if we want to respect the symmetry of the sphere, we will ask to the function $\hat{f}$ to be the radial derivative of $\hat{F}$ and to the other functions to be it's tangential derivative. In fact, provided that f has mean 0 on the sphere, there exists a $\mathcal{C}^\infty$ function $\hat{F}$, harmonic inside the ball such that $\hat{f} = \frac{\partial}{\partial\rho}\hat{F}$, where $\rho\frac{\partial}{\partial\rho} = \sum_i x^i\frac{\partial}{\partial x^i}$. Moreover, if we ask to this function $\hat{F}$ to be the harmonic prolongation inside the ball of a mean 0 function F on $\mathbf{S}^n$, then this function $\hat{F}$ is unique. We may therefore define the RIESZ transform of the function f to be the boundary value of the radial derivative of $\hat{F}$. Let us call this RIESZ transform the **analytic** one.

Let us compare those 2 definitions of the RIESZ transforms on the spheres.

Given a function f on $\mathbf{S}^n$, it's harmonic prolongation $\hat{f}$ is given by the POISSON kernel :

$$\hat{f}(x) = \int_{\mathbf{S}^n} f(y)p(x,y)dy, \tag{3}$$

(*) The pth eigenvectors are the restriction to the sphere of the homogeneous harmonic polynomials in $\mathcal{R}^{n+1}$ of degree p whose eigenvalue are $-p(p+n-1)$ (cf [CW], for further details).

where, this time, $p(x,y)$ is the POISSON kernel on the sphere : at a point $x=(r,\xi)$ of the ball, where $|x|=r$ and ξ is on the sphere, we have

$$p(x,y)=\frac{1-r^2}{\left[(1-r)^2+r|\xi-y|^2\right]^{\frac{n+1}{2}}}.$$

By definition, we know that, inside the ball, it is a solution of $\Delta\hat{f}=0$. Let us rewrite this last equation in terms of the LAPLACE operator of the sphere $\Delta_{\mathbf{S}^n}$. We get

(4) $$\Delta=(\frac{\partial}{\partial\rho})^2+\frac{n}{\rho}\frac{\partial}{\partial\rho}+\frac{1}{\rho^2}\Delta_{\mathbf{S}^n},$$

where $\frac{\partial}{\partial\rho}$ denotes as before the partial derivative with respect to the radius ρ.

No, if we set $\rho=e^{-t}$, with $t\in[0,\infty)$, one gets

(5) $$\Delta=e^{-2t}\left[(\frac{\partial}{\partial t})^2-(n-1)\frac{\partial}{\partial t}+\Delta_{\mathbf{S}^n}\right],$$

and so the harmonic prolongation $\hat{f}$ of a function f on $\mathbf{S}^n$ is in fact a solution of the equation

(6) $$\left[(\frac{\partial}{\partial t})^2-(n-1)\frac{\partial}{\partial t}+\Delta_{\mathbf{S}^n}\right]f=0\,.$$

Let us consider an operator C_n which is, in terms of the spectral decomposition of $\Delta_{\mathbf{S}^n}$, a solution of the equation

(7) $$C_n^2-(n-1)C_n+\Delta_{\mathbf{S}^n}=0\,;$$

there is a unique negative solution (negative here means that all the eigenvalues of this operator are negative) which is given by

(8) $$2C_n=(n-1)\mathbf{I}-\left[(n-1)^2-4\Delta_{\mathbf{S}^n}\right]^{1/2}\ :$$

This operator acts on an eigen vector f with eigenvalue $-p(p+n-1)$ by $C_n(f)=-pf$, while the CAUCHY operator $C=-(\Delta_{\mathbf{S}^n})^{1/2}$ which appears in the geometric RIESZ transforms is given by

$$C(f)\;=\;-\sqrt{p(p+n-1)}f.$$

So, the semigroup $\mathbf{Q}_t^n = \exp(tC_n)$ is a semigroup of bounded operators in $\mathbf{L}^2(\mathbf{S}^n)$ such that, for each function f on the sphere, the function $\tilde{f}(x,t) = \mathbf{Q}_t^n(x)$ is a solution of the equation (5) with boundary values f : by a uniqueness argument, we can now see that in fact $\tilde{f}(x,t) = \hat{f}(x, e^{-t})$, where $\hat{f}$ is the harmonic prolongation given by the POISSON kernel.

Let now f be a function with mean value 0 on the sphere, and let F be the harmonic prolongation inside the ball whose radial derivative is $\hat{f}$. From what we just saw, it is easy to check that the boundary value of F is nothing but $C_n^{-1}(f)$. (Notice that this operator is bounded on $\mathbf{L}^2(\mathbf{S}^n)$ when it acts on functions of mean 0.) Now, it is clear that the analytic RIESZ transform on the sphere is the (vector valued) operator

$$(9) \qquad \nabla(C_n^{-1}) = \nabla\{(n-1)\mathbf{I} - [(n-1)^2 - 4\Delta_{\mathbf{S}^n}]^{1/2}\}^{-1}.$$

More generally, we will call RIESZ transform on the sphere every vector-valued operator of the form
$\nabla\{\sigma\mathbf{I} - [\sigma^2 - \Delta_{\mathbf{S}^n}]^{1/2}\}^{-1}$ where σ is a positive real number : this corresponds to harmonic prolongation to the half-space which are solutions of an equation

$$[(\frac{\partial}{\partial t})^2 - 2\sigma\frac{\partial}{\partial t} + \Delta_{\mathbf{S}^n}]\hat{f} = 0\,.$$

(the reason why we restrict ourselves to positive values of σ will appear later, when we will consider the general case). Let us denote $\vec{\mathbf{R}}^\sigma$ this vector valued operator ; we have the following result :

Theorem .—For every p in $(1, \infty)$, there exists constants $C(p, \sigma, n)$ such that, for each function f on the sphere $\mathbf{S}^n$ with mean 0,

$$\frac{1}{C(p,\sigma,n)}\|f\|_p \leq \|\vec{\mathbf{R}}^\sigma(f)\|_p \leq C(p,\sigma,n)\|f\|_p.$$

Moreover, the above constant $C(p, \sigma, n)$ can be majorized by $C(p)(1 + \frac{\sigma}{n^{1/2}})$, where the constants $C(p)$ depends only on p.

If we want to work with a sphere of radius R instead of a sphere of radius 1, we get obviously the same results, but then, the constants vary like $C(p)(1+\frac{\sigma R}{n^{1/2}})$(*).

We see now that the kind of $\mathbf{H}^1$ result that we expect for the spherical RIESZ transforms will depend on the transformation under consideration ; if we are dealing with the analytic transform, we associate to a function f on the sphere it's harmonic prolongation to the ball : then we construct a martingale M_t^f by using a Brownian motion starting from 0, and stopped when it reaches the sphere.

We can compare the $\mathbf{L}^1(\mathbf{S}^n)$ norm of the RIESZ transform to the $\mathbf{H}^1$ norm of this martingale. As in $\mathcal{R}^n$, but only for function with mean 0 on the sphere, we have

$$(10) \qquad c\|f\|_{\mathbf{H}^1} \leq \|M\|_{\mathbf{H}^1} \leq C(n)\|f\|_{\mathbf{H}^1},$$

with the same meaning for the $\mathbf{H}^1$ norms than before.

Things are less clear when we consider the geometric RIESZ transform : it is related to another kind of harmonic prolongation to the space $\mathbf{S}^n \times \mathcal{R}_+$ of a function f defined on $\mathbf{S}^n$. This harmonic prolongation $\hat{f}(x,t)$ is a solution of the equation

$$[(\frac{\partial}{\partial t})^2 + \Delta_{\mathbf{S}^n}]\hat{f} = 0.$$

The associated martingale is constructed in the following way : consider first a Brownian motion X_t on the sphere, with the LEBESGUE measure of the sphere as initial distribution. This is a diffusion processus whose generator is $\Delta_{\mathbf{S}^n}$; let us also consider an auxiliary independant Brownian motion B_t^a on the real line, starting from a point $a > 0$ and stopped at the first exit time T of the positive half-line. Then, $M_t^a = \hat{f}(X_{t\wedge T}, B_{t\wedge T}^a)$ is a martingale.

For a function with mean 0 on the sphere, we get an equivalence

$$c\|f\|_{\mathbf{H}^1} \leq \lim_{a\to\infty} \|M^a\|_{\mathbf{H}^1} \leq C(n)\|f\|_{\mathbf{H}^1}.$$

In a sense, the results on the sphere seem very similar to the results on $\mathcal{R}^n$. In fact, they rely on the same structure : the spherical

---

(*) Those considerations about the behaviour of the constants in function of the different parameters will appear later when we will consider the case of the ORNSTEIN-UHLENBECK operator.

LAPLACE operator may be written as the sum $\Delta_{S^n} = \sum_{i,j}(D_{ij})^2$, where the operators D_{ij} are the restriction to the sphere of the infinitesimal rotations $x_i \frac{\partial}{\partial x^j} - x_j \frac{\partial}{\partial x^i}$. Although those operators do not commute, they commute with Δ_{S^n} : at least for the $\mathbf{L}^p(\mathbf{S}^n)$ results, this simple fact makes the proof of MEYER work perfectly. In fact, this previous property of the sphere is shared by all the homogeneous spaces (ratio of two compact LIE groups), and this allows one to extend a lot of these results to the case of homogeneous spaces.

d)—The RIESZ transforms for the ultraspherical operators and the ORNSTEIN-UHLENBECK case.

Now, let us project our sphere on one of it's diameters : this means that we restrict ourselves to functions $f(x)$ on the sphere wich depend only the projection of x onto one diameter. Let us call x_1 this projection: it is a point in the interval $[-1,1]$. If $f(x_1)$ is such a function (we will call them *zonal functions*), then $\Delta_{S^n}(f)$ is also zonal, and we have

$$\text{(11)} \qquad \Delta_{S^n}(f)(x_1) = \mathbf{L}_n(f)(x_1) = (1-x_1^2)\frac{\partial^2 f}{\partial x_1^2} - n x_1 \frac{\partial f}{\partial x_1}.$$

This $\mathbf{L}_n$ is a differential operator on the interval. It is symmetric with respect to the projection of the LEBESGUE measure of the sphere onto the diameter, which is $\mu_n(dx_1) = (1-x_1^2)^{\frac{n}{2}-1}dx_1$. If $n \geq 2$, this operator is even self-adjoint negative with respect to this measure, and we may define as before the negative operators $C_n^\sigma = \sigma\mathbf{I} - \sqrt{\sigma^2\mathbf{I} - \mathbf{L}_n}$. For a zonal function, we have of course $[\sigma\mathbf{I} - \sqrt{\sigma^2\mathbf{I} - \Delta_{S^n}}]f(x_1) = C_n^\sigma f(x_1)$.

We also have $|\nabla f|^2 = (\frac{\partial f}{\partial x_1})^2$, so that the results about the RIESZ transforms that we got for the sphere carry out for these operators : the operators $\frac{\partial}{\partial t} \circ C_n^\sigma$ are bounded on the subspace $\mathbf{L}^p(\mu_n)$ consisting of functions with mean value 0.

We may replace in the above formula the variable n by any real number ν, and we get then the *ultraspherical operator* of order ν. This operator is a symmetric operator with respect to the measure $\mu_\nu(dx)$, and it is self-adjoint when $\nu \geq 2$. Are the results on RIESZ transforms still valid when ν is not an integer? As we may guess, the answer is yes, but we must notice that the homogeneous space structure of the sphere which enabled to proove the results is no longer valid in this context : there is no first order operator on the interval $(-1, 1)$ which commutes with the operator $\mathbf{L}_\nu$.

Let us rescale the operator $\mathbf{L}_n$ in order to get an operator on $(-n^{1/2}, n^{1/2})$: this means that we are now projecting the sphere of radius $n^{1/2}$ instead of the sphere of radius 1. We get the operator $\widehat{\mathbf{L}_n} f(x) = (1 - \frac{x^2}{n})\frac{\partial^2 f}{\partial x^2} - x\frac{\partial f}{\partial x}$. If we let n converge to infinity in the above formula, we see that this operator converges, at least in a week sense, to the ORNSTEIN-UHLENBECK operator $\mathbf{L}f(x) = \frac{\partial^2 f}{\partial x^2} - x\frac{\partial f}{\partial x}$. This operator is a self-adjoint operator with respect to the Gaussian measure on the real line. It is also negative, and we may define it's associated RIESZ transforms. For simplicity, let us restrict ourselves to the simplest one : $R = \frac{\partial}{\partial t} \circ (-\mathbf{L})^{-1/2}$. MEYER [M2] prooved that this operator is bounded on the space $\mathbf{L}^p_0$ of the Gaussian measure, for $1 < p < \infty$. He even prooved that there is a norm equivalence

$$c(p)\|df\|_p \leq \|(-\mathbf{L})^{1/2} f\|_p \leq C(p)\|f\|_p.$$

This estimate carries out for the ORNSTEIN-UHLENBECK operator in $\mathcal{R}^n$, $\mathbf{L} = \Delta - \sum_i x^i \frac{\partial}{\partial x^i}$, wich is symmetric with respect to the n-dimensionnal Gaussian measure, **with constants independant on the dimension** (this is an important point in MALLIAVIN calculus, because we may then get this estimate on the WIENER space). The important point in MEYER's proof is that the second order part of the ORNSTEIN-UHLENBECK operator is $\sum_i (\frac{\partial}{\partial x^i})^2$, with a very simple commutator $[\mathbf{L}, \frac{\partial}{\partial x^i}] = \frac{\partial}{\partial x^i}$. But this slight difference with the classical cases, where this commutator is 0, introduces a very strong complication in the proof of result on RIESZ transforms.

So we can see that the $\mathbf{L}^p$ theory for the ultraspherical operators and for ORNSTEIN-UHLENBECK are quite similar : in fact, one could certainly deduce the ORNSTEIN-UHLENBECK case from the spherical one, by taking limits when the dimension tends to infinity. But if we consider the $\mathbf{H}^1$ theory, then the things become different : the $\mathbf{H}^1$ theory depends heavily on the dimension, and, even in dimension 1, the ORNSTEIN-UHLENBECK operator may be considered as an infinite dimensionnal operator.

To be more precise, let us describe the simplest case of $\mathbf{H}^1$ theory for the ultraspherical operators : fix an index ν and consider a function f on the interval$(-1,1)$ with mean 0 with respect to the measure $\mu_\nu(dx)$. We consider it's "harmonic" prolongation $\hat{f}$ to the half-space $(-1,1) \times \mathcal{R}_+$, it is a solution of the equation

$$[\frac{\partial^2}{\partial t^2} + \mathbf{L}_\nu](\hat{f}) = 0.$$

This solution is given by $\hat{f}(x,t) = \exp(-t(-\mathbf{L}_\nu)^{1/2})f(x)$. Then, as usual, we may associate a martingale to this solution : let X_t be the process on $(-1,1)$ whose infinitesimal generator is $\mathbf{L}_\nu$ (*): if ν is an integer, this is just the projection of the spherical Brownian motion onto the diameter. To get the usual martingale, we as before pick an auxiliary Brownian motion B_t^a starting from $a > 0$ and consider the process $\hat{f}(X_t, B_t^a)$. Now, we get the usual results comparing the probalistic $\mathbf{H}^1$ norm and the norm that we get when we consider the RIESZ transform in $\mathbf{L}^1$.

The $\mathbf{H}^1$ theory for the ultraspherical operator of order ν relies on the following subharmonicity property : let $\hat{f}(x,t)$ be a solution of the equation $[\frac{\partial^2}{\partial t^2} + \mathbf{L}_\nu](\hat{f}) = 0$ in an open set of $\mathcal{R}^2$, and let $|\nabla f|^2$ denotes $(\frac{\partial}{\partial t}f)^2 + (\frac{\partial}{\partial x}f)^2$. Then

$$\mathbf{L}_\nu(|\nabla f|^p \geq 0, \quad \text{as soon as} \quad p \geq \frac{\nu}{\nu-1}.$$

If we compare this property to the subharmonicity property in $\mathcal{R}^n$, we see that this parameter ν in the ultraspherical operator behaves like a dimension (as we expected from the definition).

This subharmonicity property of $\mathbf{L}_\nu$ is shared by every operator on an interval of $\mathcal{R}$ with can be written (up to a change of variable)

(*) We restrict ourselves to the case where $\nu > 1$; if $\nu \geq 2$, then this process never reaches the boundary; if $1 < \nu < 2$, then we take the reflected process at the boundary (*cf* [B2])

as $\dfrac{\partial^2}{\partial x^2} - a(x)\dfrac{\partial}{\partial x}$, where the function $a(x)$ satisfy the differential inequality $\dfrac{\partial}{\partial x}a \geq \dfrac{a^2}{\nu - 1}$. For the ORNSTEIN-UHLENBECK operator, we have $a(x) = x$, and this inequality is not satisfied. This explains why there is no subharmonicity property for ORNSTEIN-UHLENBECK (which behaves like an infinite dimensionnal operator) and why, at least for the moment, there is no $\mathbf{H}^1$ theory for it.

2.—The general case.

All the different results about RIESZ transforms that we saw in the introduction concerned vector valued operators : to be more precise, they in fact deal with 1-form valued operators, as we will show in the following. These results can be restated in the general context of symmetric second order operators on manifolds, which we will describe now. We first start with some basic vocabulary and notations of differential geometry, which are certainly familiar to the reader.

From now on, we shall work on a p-dimensionnal manifold $\mathbf{E}$, $\mathcal{C}^\infty$ and connected. We will denote by $\mathcal{C}_c^\infty$ the space of functions on $\mathbf{E}$ which are in the $\mathcal{C}^\infty$ class and have compact support. We will denote by $(x^i,\ 1 \leq i \leq p)$ a local system of ($\mathcal{C}^\infty$) coordinates.

Notation.

We shall also adopt the summation convention over the repeated indices : this means that whenever the same index appears up and down in an expression, it is understood that we sum over it. For example, the expression $\omega_i X^i$ stands for $\sum_i \omega_i X^i$, etc.

Vector fields, 1-forms and more general tensors.

In a system of coordinates, we will consider vector fields $X = (X^i(x))$ and (1-)forms $\omega = (\omega_i(x))$. A vector field is naturally associated with a first order operator acting on $\mathcal{C}_c^\infty$ by $X(f) = X^i\dfrac{\partial}{\partial x^i}(f)$. We will always restrict our attention to smooth vector fields, i.e. those who have $\mathcal{C}^\infty$ coefficients in a system of coordinates. When we consider a new system of coordinates $y^j(x^1, \cdots, x^p)$, the coordinates of the vector field X change according to the rule

$$X^j(y) = \frac{\partial y^j}{\partial x^i} X^i(x).$$

In our system of coordinates, a basis of the 1-form is given by $(dx^i, 1 \le i \le p)$. The general 1-form is given in this basis by it's ($\mathcal{C}^\infty$) components $(\omega_i) : \omega = \omega_i dx^i$. We do not need any formal definition of 1-forms, and it is enough to know that they act on vector fields via the duality coupling $\langle \omega, X \rangle = \omega_i X^i$. Also, the main 1-forms (but not the only ones) that we will consider in the following are constructed from functions : $df = \frac{\partial}{\partial x^i} f dx^i$. In a change of variable, a 1-form behaves in the opposite way than a vector field :

$$\omega_j(y) = \frac{\partial x^i}{\partial y^j} \omega_i(x).$$

Later on, we will consider more general tensors than just vector fields and 1-forms. A tensor T is represented in a system of coordinates by a family of coordinates $T^{i_1 \cdots i_{p_1}}{}_{i_{p_1+1} \cdots i_{p_2}} \cdots {}^{i_{p_{k-1}+1} \cdots i_{p_k}}(x)$; in compact notations, we will denote this set of indices by $\mathbf{I}$ (this takes in account the position of the indices, down or up) ; also the coordinates of the tensor are denoted $T^{\mathbf{I}}$, even if these indices are down and not up. The charasteristic property of a tensor is the following : if we change variables, we have to multiply the coordinates of T by the matrix $(\frac{\partial y^i}{\partial x^j})$ for each index which is up and by the matrix $(\frac{\partial x^i}{\partial y^j})$ for each index which is down. This gives an awfull formula

$$T^{j_1 \cdots j_{p_1}}{}_{j_{p_1+1} \cdots j_{p_2}} \cdots {}^{j_{p_{k-1}+1} \cdots j_{p_k}}(y) =$$
$$\frac{\partial y^{j_1}}{\partial x^{i_1}} \cdots \frac{\partial y^{j_{p_1}}}{\partial x^{i_{p_1}}} \cdots \frac{\partial y^{j_{p_k}}}{\partial x^{i_{p_k}}} T^{i_1 \cdots i_{p_1}}{}_{i_{p_1+1} \cdots i_{p_2}} \cdots {}^{i_{p_{k-1}+1} \cdots i_{p_k}}(x).$$

Of course, we will never use such a formula and the worst tensors we will encounter will just have 2 indices.

Tensors with many indices are in general constructed from simpler tensors by tensor products : if we have a tensor T with a set of indices $\mathbf{I}$ and coordinates $T^{\mathbf{I}}$ and another tensor S with coordinates $\mathbf{J}$, we construct a new tensor $T \otimes S$ with set of coordinates $\mathbf{I} \cup \mathbf{J}$ by setting $(T \otimes S)^{\mathbf{I} \cup \mathbf{J}} = T^{\mathbf{I}} S^{\mathbf{J}}$.

Connections.

We saw that differentiating a function in a local system of coordinates gives rise to a 1-form, wich is the basic example of tensor. If we want to repeat this operation with a 1-form, we get into trouble : if (ω_i) are the component of a 1-form, $(\frac{\partial}{\partial x^j}\omega_i)$ are no longer the components of a tensor. We have the same trouble if we want to differentiate the components of a vector field. This is why we have to introduce the notion of connection : this is a way to take derivatives of vector fields (and in fact of all tensors) such that the result is again a tensor. In general, we denote connections by the letter ∇, and we proceed as follows : if X is a vector field, then (∇X) is a tensor with two indices, with coordinates $(\nabla X)_i{}^j = \nabla_i X^j$ given by

$$\nabla_i X^j = \frac{\partial}{\partial x^i} X^j + \Gamma^j_{ik} X^k,$$

where the coefficients Γ^j_{ik} are called the CHRISTOFFEL symbols of the connection. They are not the components of a tensor, and they must satisfy a specific change of variable formula : if $\Gamma^j_{ik}(x)$ (resp. $\Gamma^j_{ik}(y)$) are the symbols of ∇ in the system of coordinates $x = (x^i)$ (resp. $y = (y^j)$), then we have

$$\Gamma^j_{ir}(y) = \frac{\partial y^j}{\partial x^q}\frac{\partial x^p}{\partial y^i}\frac{\partial x^l}{\partial y^r}\Gamma^q_{pl}(x) + \frac{\partial y^j}{\partial x^q}\frac{\partial^2 x^q}{\partial y^i \partial y^r}.$$

Once we know how to take derivatives of a vector field , we know how to take derivatives of any kind of tensor ; if T if a tensor with set of indices $\mathbf{I}$, then ∇T is a tensor with one more index, always down and coming first. If $T^{\mathbf{I}}$ are the coordinates of T, then ∇T has coordinates $\nabla_i T^{\mathbf{I}}$. The rules are the following :

a) If ω is a 1-form with coordinates (ω_i), we set $\nabla_i \omega_j = \frac{\partial}{\partial x^i}\omega_j - \Gamma^k_{ij}\omega_k$, such that, for each vector field X and every 1-form ω, we have

$$d\langle \omega, X\rangle_i = (\nabla_i X^l)\omega_l + (\nabla_i \omega_l) X^l.$$

b) If we have 2 tensors T_1 and T_2, we have

$$\nabla(T_1 \otimes T_2) = (\nabla T_1)\otimes T_2 + T_1 \otimes (\nabla T_2).$$

With these rules, we have a consistent set of notations, and we may write $\nabla_i f$, for example, instead of $(df)_i$. Unfortunately, unlike the usual calculus, it is not true that, for each function f, the tensor $(\nabla_i \nabla_j f)$ is symmetric. If such is the case, we shall say that the connection is **torsion free.** In the following, we will restrict our attention to such connections.

Even if ∇ is torsion free, it is not true in general that, for a given vector field X, the tensor $(\nabla_i \nabla_j X^k)$ is symmetric in the indices i and j. There is a tensor $(R_{ij}{}^k{}_l)$, called the **curvature tensor of the connection** ∇, such that, for each vector field X, one has $\nabla_i \nabla_j X^k - \nabla_j \nabla_i X^k = R_{ij}{}^k{}_l X^l$. The RICCI **tensor of the connection** ∇ is the tensor $\rho_{il} = R_{ji}{}^j{}_l$.

Elliptic second order differential operators.

A elliptic second order differential operator **L** on **E** is given in a local system of coordinates by

$$\mathbf{L}f(x) = g^{ij}(x)\frac{\partial^2}{\partial x^j x^j} f(x) + b^i(x)\frac{\partial}{\partial x^i} f(x),$$

where the coefficients g^{ij} and b^i are $\mathcal{C}^\infty$, and where (g^{ij}) is a definite positive symmetric matrix. It appears that the matrix (g^{ij}) is a tensor (i.e. satisfy the change of variables rule) but the coefficient (b^i) do not form a vector field. So, if we want to restrict our attention to those objects which are invariant under change of coordinates, we will have to look at the **canonical decomposition of L**, which we will describe now.

Notations.

From now on, we will adopt the following conventions : we will denote (g_{ij}) the inverse matrix of the symmetric positive matrix (g^{ij}) (this gives us a tensor which is sometimes called the Riemannian metric associated to **L**).
When we have a vector field X with coordinates (X^i), we may "lower" it's index according to this matrix : this means that we associate to X a 1-form X_* whose coordinates are $X_i = g_{ij}X^j$ (We omit the * when the context is clear). On the other hand, if we have a 1-form ω with coordinates ω_i, we can "lift" it's index with the matrix g^{ij} to get a vector field whose coordinates are $\omega^i = g^{ij}\omega_j$. This operation of lifting and lowering indices allows us to indentify vector fields and 1-forms.

In the same way, we may lift or lower whatever index we want in a general tensor. As we will see in a moment, this operation allows us to give more compact formulas.

Scalar products of tensors.

The metric (g) gives us a scalar product on vectors by the formula $X.Y = X^iY_i$, where $X = X^i\frac{\partial}{\partial x^i}$ and $Y = Y^i\frac{\partial}{\partial x^i}$. In the same way, it also gives us a scalar product on 1-forms through the just described indentification of vectors an 1-forms.

This scalar product extends to all kind of tensors via the formula $(T\otimes S).(T'\otimes S') = (T.T')(S\otimes S')$, where the tensors T and T' (resp. S and S') have the same type. For example, for 2-tensors $T = (T^{ij})$ and $S = (S^{ij})$, we have $T.S = T^{ij}S_{ij}$. In general, we will denote by $|T|$ the norm of a tensor : $|T|^2 = T.T$. **We will use the same definition for all kind of tensors, including the p-forms.**

Canonical decomposition.

Associated to the Riemannian metric g is a **unique Riemannian connection** ∇ : this is the unique connection which is torsion free and such that $\nabla g = 0$, for the tensor g^{ij} (or equivalently for the tensor g_{ij}). In a local system of coordinates, the CHRISTOFFEL symbols of the connection are

$$\Gamma^i_{jk} = \frac{1}{2}g^{ip}(\frac{\partial}{\partial x^k}g_{pl} + \frac{\partial}{\partial x^l}g_{pk} - \frac{\partial}{\partial x^p}g_{kj}).$$

Now, we may consider the LAPLACE-BELTRAMI operator associated with the metric g : it is the operator given in a local system of coordinates by

$$\Delta f = g^{ij}\nabla_i\nabla_j f = \nabla^i\nabla_i f = \nabla_i\nabla^i f.$$

Since we are only dealing with tensors in this expression, this definition is independant of the choice of the coordinate system. The difference between Δ and $\mathbf{L}$ is a first order operator, i.e. a vector field X. **This decomposition $\mathbf{L} = \Delta + X$ is called the canonical decomposition of $\mathbf{L}$.**

The Ricci curvature of a second order differential operator.

Let us consider the Ricci tensor ρ of the connection ∇ : it is a symmetric tensor, that is, in a local system of coordinates, we have $\rho_{ij} = \rho_{ji}$. By definition, the Ricci tensor of Δ will be the tensor whose components are $\mathrm{Ric}(\Delta)^{ij} = \rho^{ij}$. Since it is a symmetric tensor, we may identify it with a symmetric bilinear operator maping 1-forms into functions : $\mathrm{Ric}(\Delta)(\omega, \eta) = \rho^{ij}\omega_i\eta_j$.

Let us also denote $\nabla^s X$ the symmetric tensor given in a system of coordinates by $\nabla^s X^{ij} = \frac{1}{2}(\nabla^i X^j + \nabla^j X^i)$, i.e. the tensor that we get in we symmetrise the tensor ∇X after lifting its indices. Then we set

Definition.—*The* Ricci *tensor of* $\mathbf{L}$ *is the tensor* $\mathrm{Ric}(\Delta) - \nabla^s X$ *: we will denote it* $\mathrm{Ric}(\mathbf{L})$.

If order to understand why this tensor will play a important role in what follows, let us introduce some new notions. Consider a bilinear operation K maping a pair of functions (f, g) into a new function $K(f, g)$, and let us assume that this operation is symmetric in f and g. With the help of the operator $\mathbf{L}$, we can construct a new symmetric bilinear operator $\partial_{\mathbf{L}} K$ by the following formula :

$$2\partial_{\mathbf{L}} K(f, g) = \mathbf{L}\big(K(f, g)\big) - K(\mathbf{L}f, g) - K(f, \mathbf{L}g).$$

Let us start with the simplest such operator : $\Gamma_0(f, g) = fg$. Then, we get

$$\partial_{\mathbf{L}} \Gamma_0(f, g) = \Gamma_1(f, g) = \nabla^i f \nabla_i g.$$

In what follows, we will often write this $\Gamma(f, g)$ or simply $\nabla f . \nabla g$. Note that, in the canonical decomposition of $\mathbf{L}$, it does not depend of the vector field X. If we go on, the second operator we get is the following

$$\partial_{\mathbf{L}} \Gamma_1(f, g) = \Gamma_2(f, g) = \nabla^i \nabla j f \nabla^i \nabla^j g + \mathrm{Ric}(\mathbf{L})(df, dg).$$

This comes from a straightforward computation in a local system of coordinates (*cf* [B3], for example). In the case where there is no vector field X, this formula is known as the Bochner-Lichnerowicz-Weitzenbock formula.

Now, the Ricci tensor of $\mathbf{L}$ appears as the largest bilinear symmetric operator R on 1-forms such that, for each function f, one has $\Gamma_2(f, f) \geq R(df, df)$.

Let us consider the lowest eigenvalue of the tensor Ric(**L**) in the metric g : this means that, in a local system of coordinates, we consider the lowest eigenvalue $r(x)$ of the matrix $\mathrm{Ric}(\mathbf{L})^i_j$. (This is of course independant of the choice of the coordinates.) This function $r(x)$ is the largest function satisfying the following inequality :

$$\forall f, \quad \Gamma_2(f,f) \geq r(x)\Gamma(f,f).$$

Definition.—*We will say that the* RICCI *curvature of* **L** *is bounded from below (resp. by a constant* λ*) iff this function* $r(x)$ *is bounded from below (resp.* $r(x) \geq \lambda$*).*

Symmetry and self-adjointness.

Let dx denote the RIEMANN measure associated with the metric g : in a local system of coordinates

$$dx = [det(g_{ij})]^{1/2} dx^1 \cdots dx^p. \qquad \text{Then,}$$

$$\forall f, g \in \mathcal{C}_c^\infty, \quad \int f\Delta g\, dx = \int g\Delta f\, dx = -\int \nabla f . \nabla g\, dx.$$

From this, a short computation shows that **L** is symmetric to a measure $\mu(dx)$ with density $e^{h(x)}$ with respect to the measure dx iff the vector field X which appears in the canonical decomposition is equal to ∇h. Then, we have

$$\forall f, g \in \mathcal{C}_c^\infty, \quad \int f\mathbf{L}g\, \mu(dx) = \int g\mathbf{L}f\, \mu(dx) = -\int \nabla f . \nabla g\, \mu(dx).$$

Convention and notation.

From now on, we will assume that $X = \nabla h$: since the function h is defined up to an additive constant, we may always assume that the measure μ has either mass 1, either infinity, depending on e^h being integrable or not. We will denote by $\langle f \rangle$ the integral $\int f(x)\, \mu(dx)$ and by $\langle f, g \rangle$ the scalar product in $\mathbf{L}^2(\mu)$: $\langle f, g \rangle = \int fg\, d\mu$.

Since we always have $\langle f, \mathbf{L}f \rangle \leq 0$, for every f in $\mathcal{C}_c^\infty$, we know that **L** has a self adjoint extension. But this extension is not unique in general, and the description of the operator $\mathbf{L}$ on $\mathcal{C}_c^\infty$ is not enough to describe this extension. This is why we will add to our assumptions the following hypothesis :

Hypothesis : **The manifold E is complete for the Riemannian structure g.**

This assumption is equivalent to the following : There exists a sequence f_n in $\mathcal{C}_c^\infty$ such that

$$0 \leq f_n \leq f_{n+1} \leq 1 \,; \quad f_n \to 1\,(n \to \infty)\,; \quad \nabla f_n . \nabla f_n \leq \frac{1}{n}.$$

When this is the case, then the self adjoint extension of $\mathbf{L}$ is unique, and $\mathcal{C}_c^\infty$ is dense in the $\mathbf{L}^2(\mu)$-domain of this extension : we say that $\mathbf{L}$ is essentially self adjoint. (*cf* [B4] or [Str] for example). We have a spectral decomposition

$$\mathbf{L} = -\int_0^\infty \lambda \, dE_\lambda.$$

When the manifold is compact, this reduces to a decomposition of $\mathbf{L}^2(\mu)$ into an orthogonal sum $\mathbf{L}^2(\mu) = \oplus_n E_n$, such that each E_n is an finite dimensionnal eigenspace of $\mathbf{L}$ with eigenvalue $-\lambda_n \leq 0$.

Heat semigroup and harmonic prolongations.

The heat semigroup associated with $\mathbf{L}$ is the semigroup

$$P_t = \int_0^\infty e^{-\lambda t} \, dE_\lambda = e^{t\mathbf{L}}.$$

It satisfies the following properties :

a) Each operator P_t is self adjoint ;

b) It is represented by a kernel : $P_t(f)(x) = \int f(y)\, p_t(x,y)\, \mu(dy)$, where the functions $p_t(x,y)$ are positive, symmetric in (x,y), and smooth in the variables (t,x,y) in the domain $(t>0)$;

c) It is submarkovian : $P_t(1) = \int p_t(x,y)\, \mu(dy) \leq 1$;

d) The operators P_t form a semigroup of contractions in each space $\mathbf{L}^p(\mu)$ $(1 \leq p < \infty)$. In particular, for each f in $\mathbf{L}^p(\mu)$, $P_t(f) \to f$ when $t \to 0$.

Moreover, if Ric(**L**) is bounded from below, then it is Markovian : $P_t(1) = 1$. (*cf* [B5], for example.) From now on, we will assume that this last condition holds.

Let us now introduce the operators $C^\sigma = \sigma I - \sqrt{\sigma^2 I - L}$ that we considered in the first part in connection with the RIESZ transforms on the spheres. When $\sigma \geq 0$, they are also the infinitesimal generators of Markovian semigroups $Q_t^\sigma = e^{tC^\sigma}$. To see that, it is enough to remark that we have a subordination formula

$$Q_t^\sigma = \int_0^\infty P_s\, h(t,s,\sigma)\, ds \quad \text{with}$$

$$h(t,s,\sigma) = \pi^{-1/2} t s^{-3/2} e^{-t^2/4s} e^{\sigma t - \sigma^2 s}.(*)$$

This function $h(t,s,\sigma)$ is such that

$$\int_0^\infty h(t,s,\sigma)\, e^{-\lambda s}\, ds = e^{t(\sigma - \sqrt{\sigma^2+\lambda})},$$

so that the measure $\mu_t^\sigma(ds) = h(t,s,\sigma)ds$ form for each positive σ a convolution semigroup of probability measures on $\mathcal{R}_+$.

The operators C^σ satisfy the identity

$$(C^\sigma)^2 - 2\sigma C^\sigma + \mathbf{L} = 0.$$

Hence, if f is a bounded integrable function on **E**, the function $\hat{f}(x,t) = Q_t^\sigma(f)(x)$ is a solution of the equation $(\frac{\partial^2}{\partial t^2} - 2\sigma\frac{\partial}{\partial t} + \mathbf{L})\hat{f} = 0$. Since this last operator is elliptic on $\mathbf{E} \times (0,\infty)$, the function $\hat{f}$ is $\mathcal{C}^\infty$. Moreover, if f is in $\mathcal{C}_c^\infty$, then $\hat{f}(x,t)$ converges to $f(x)$ when $t \to 0$, and this remains the case for almost every x as soon as f is measurable bounded and integrable.

The extension of L to the 1-forms.

Together with the operator **L** acting on functions, we will consider an operator $\vec{\mathbf{L}}$ acting on 1-forms and satisfying the following properties

a) $\forall f \in \mathcal{C}_c^\infty$, $\vec{\mathbf{L}}(df) = d(\mathbf{L}f)$;

(*) When $\sigma < 0$, the same formula would lead to $Q_t^\sigma(1) = e^{-2\sigma t}$.

b) $\forall \omega \in \mathcal{C}_c^\infty$, $\mathbf{L}|\omega|^2 = 2\omega.\vec{\mathbf{L}}\omega + 2|\nabla\omega|^2 + 2R(\omega,\omega)$, where R is a symmetric tensor (R^{ij}), such that, for every 1-form $\omega = \omega_i dx^i$, we have $R(\omega,\omega) = R^{ij}\omega_i\omega_j$.

Such an operator exists and is unique : the tensor R is then the RICCI tensor $\mathrm{Ric}(\mathbf{L})$, and , in a local system of coordinates, we have

$$(\vec{\mathbf{L}}\omega)_i = \nabla_j\nabla^j\omega_i + \vec{X}(\omega)_i - \mathrm{Ric}(\mathbf{L})_i^j\omega_j,$$

where $\vec{X}$ is the "horizontal lifting" of the vector field X which appears in the canonical decomposition of $\mathbf{L}$, i.e.

$$\vec{X}(\omega)_i = (\nabla_i X^j)\omega_j.$$

In our context, the operator $\vec{\mathbf{L}}$ will play the role of the usual DE RHAM operator of Riemannian geometry. In order to see this, let us introduce the space $\vec{\mathbf{L}}^2(\mu)$ of 1-forms in $\mathbf{L}^2(\mu)$: this is the completion of the space of 1-forms in $\mathcal{C}_c^\infty$ with the norm $\|\omega\|_2^2 = \langle|\omega|^2\rangle$. For a 1-form $\omega = \omega_i dx^i$, we denote by $d\omega$ the 2-form whose coordinates are $\frac{1}{\sqrt{2}}(\frac{\partial}{\partial x^i}\omega_j - \frac{\partial}{\partial x^j}\omega_i)$(*).Then, we may introduce the operator ∂ mapping 1-forms on functions and 2-forms on 1-forms by the following:

For 1-forms : for every 1-form ω and every function f in $\mathcal{C}_c^\infty$, we have $\langle\partial\omega.f\rangle = \langle\omega.df\rangle$.

For 2-forms: for every 2-form η and every 1-form ω in $\mathcal{C}_c^\infty$, we have $\langle\partial\eta.\omega\rangle = \langle\eta.d\omega\rangle$. Then, a short computation in a local system of coordinates shows that

$$\vec{\mathbf{L}} = -(d\partial + \partial d). \tag{12}$$

As a consequence, for every pair of 1-forms in $\mathcal{C}_c^\infty$,we have

$$\langle\omega.\vec{\mathbf{L}}\eta\rangle = -\langle d\omega.d\eta\rangle - \langle\partial\omega.\partial\eta\rangle.$$

From this, it follows that the operator $\vec{\mathbf{L}}$ is symmetric and negative on the space $\vec{\mathbf{L}}^2(\mu)$; in fact, since we have assumed that the space

(*) This strange factor $1/\sqrt{2}$ comes from our conventions on the norm of a 2-form, which is unusual in geometry.

E is complete, the operator $\vec{\mathbf{L}}$ is (essentially) self-adjoint (*cf* [B4], for example). We therefore have a spectral decomposition in $\vec{\mathbf{L}}^2(\mu)$

$$\vec{\mathbf{L}} = -\int_0^\infty \lambda \, d\vec{E}_\lambda.$$

The operator $\vec{\mathbf{L}}$ generates a heat semigroup $\vec{\mathbf{P}}_t = \exp(t\vec{\mathbf{L}})$ of contractions in $\vec{\mathbf{L}}^2(\mu)$, and we may also consider the subordinated semigroups of contractions $\vec{\mathbf{Q}}_t^\sigma = \int_0^\infty \vec{\mathbf{P}}_s h(t,s,\sigma)\, ds$, whose generators are the operators $\vec{\mathbf{C}}^\sigma = \sigma I - \sqrt{\sigma^2 I - \vec{\mathbf{L}}}$.

For a given 1-form ω in C_c^∞, the 1-parameter family of 1-forms $\hat{\omega}(x,t) = \vec{\mathbf{Q}}_t^\sigma(\omega)(x)$ is a solution of the equation

$$(\frac{\partial^2}{\partial t^2} - 2\sigma\frac{\partial}{\partial t} + \vec{\mathbf{L}})(\hat{\omega}) = 0.$$

Therefore, because of the ellipticity assumption on $\mathbf{L}$, this family is smooth (as a solution of an elliptic differential equation). Since the relation $d\mathbf{L}(f) = \vec{\mathbf{L}}(df)$ holds for the functions f which are in C_c^∞ and because of the essential self-adjointness, we also have $d\mathbf{Q}_t^\sigma = \vec{\mathbf{Q}}_t^\sigma d$, and also $d\mathbf{C}^\sigma = \vec{\mathbf{C}}^\sigma d$.

The most important property of the Ricci curvature of the operator $\mathbf{L}$ appears in the following : if we assume that the Ricci curvature of $\mathbf{L}$ is bounded from below by a constant ρ, we have (*cf* [B4])

$$\forall \omega \in C_c^\infty,\ |\vec{\mathbf{P}}_t\omega| \le e^{-\rho t}\mathbf{P}_t|\omega|. \tag{13}$$

This behaviour of the semigroup $\vec{\mathbf{P}}_t$ reflects in the semigroups $\vec{\mathbf{Q}}_t^\sigma$ in the following way :

Proposition .— Choose σ such that $\sigma^2 + \rho \ge 0$ and set $\sigma_1^2 = \sigma^2 + \rho$. Then we have

$$|\vec{\mathbf{Q}}_t^\sigma \omega| \le e^{t(\sigma - \sigma_1)}\mathbf{Q}_t^{\sigma_1}|\omega|.$$

In particular, we have

$$\|\vec{\mathbf{Q}}_t^\sigma \omega\|_\infty \le e^{t(\sigma-\sigma_1)}\|\omega\|_\infty.$$

Proof. We write $|\vec{\mathbf{Q}}_t^\sigma \omega| \leq \int_0^\infty |\vec{\mathbf{P}}_t\omega| h(t,\sigma,s) ds$. By the above majorization of $|\vec{\mathbf{P}}_t\omega|$, we get

$$|\vec{\mathbf{Q}}_t^\sigma \omega| \leq \int_0^\infty e^{-\rho s} \mathbf{P}_t|\omega| h(t,\sigma,s) ds.$$

Now, the fundamental property of the function $h(t,\sigma,s)$ tells us that

$$\int_0^\infty e^{-\rho s} \exp(t\mathbf{L})\, h(t,\sigma,s)\, ds = \exp\{t(\sigma - \sqrt{\sigma^2 + \rho - \mathbf{L}})\}$$
$$= \exp\{t(\sigma - \sigma_1)\} \mathbf{Q}_t^{\sigma_1}.$$

∎

In the same way, we could proove that, for every $\alpha \in [1, \infty)$, we have

$$|\vec{\mathbf{Q}}_t^\sigma \omega|^\alpha \leq \exp\{t(\sigma - \sigma_\alpha)\} \mathbf{Q}_t^{\sigma_\alpha}(|\omega|^\alpha),$$

with $\sigma_\alpha = \sqrt{\sigma^2 + \alpha\rho}$, when $\sigma^2 + \alpha\rho \geq 0$. In fact, there is a more general statement about this changes of coefficient in the semigroups $\mathbf{Q}_t^\sigma$ and the exponential factors wich appear in these formulas. In order to simplify the following, we introduce a new notation :

Notation : from now on, we denote by $\mathbf{L}^\sigma$ the operator $\frac{\partial^2}{\partial t^2} - 2\sigma \frac{\partial}{\partial t} + \mathbf{L}$ on $\mathbf{E} \times \mathcal{R}_+$. In the same way, $\vec{\mathbf{L}}^\sigma$ denotes $\frac{\partial^2}{\partial t^2} - 2\sigma \frac{\partial}{\partial t} + \vec{\mathbf{L}}$.

The first remark is that, since the semigroups $\mathbf{Q}_t^\sigma$ are Markovian, we know that, if $f(x,t)$ is a bounded function with 2 continuous derivatives on $\mathbf{E} \times \mathcal{R}_+$ which satisfies $\mathbf{L}^\sigma f \geq 0$, then $f(x,t) \leq \mathbf{Q}_t^\sigma(f(\cdot, 0)$ (*cf* [B4], for example). Then, we have the following lemma

Lemma .—

1) Let $g(x,t)$ be a $\mathcal{C}^\infty$ function on $\mathbf{E} \times \mathcal{R}_+$ such that $\mathbf{L}^{\sigma_1} f = 0$. then we have

a) For every α real and for every $\sigma_2 \geq 0$

$$e^{-\alpha s} \mathbf{L}^{\sigma_2} e^{\alpha s} g^2 = 2|dg|^2 + 2|(\frac{\partial}{\partial t} + (\sigma_1 + \alpha - \sigma_2)\mathbf{I})g|^2$$
$$+ (-\alpha^2 + 2\alpha(\sigma_2 - 2\sigma_1) - 2(\sigma_1 - \sigma_2)^2) g^2.$$

b) If α is in the interval $[-\sigma_1, 0]$, then $\mathbf{L}^{\sigma_1+\alpha} e^{\alpha s}|g| \geq 0$.

2) Let $\omega(x,t)$ be a smooth family of 1-forms satisfying $\vec{\mathbf{L}}^{\sigma_1}\omega = 0$. Then we have

a) For every α real and for every $\sigma_2 \geq 0$

$$\begin{aligned} e^{-\alpha s}\mathbf{L}^{\sigma_2} e^{\alpha s}|\omega|^2 \geq \\ 2|\nabla\omega|^2 + 2|(\frac{\partial}{\partial t} + (\sigma_1 + \alpha - \sigma_2)\mathbf{I})\omega|^2 \\ +(-\alpha^2 + 2\alpha(\sigma_2 - 2\sigma_1) - 2(\sigma_1 - \sigma_2)^2 + 2\rho)|\omega|^2. \end{aligned}$$

b) If $\sigma_1^2 + \rho \geq 0$ and if α is in the interval $[-\sigma_1, -\sigma_1 + \sqrt{\sigma_1^2 + \rho}]$, then with $\sigma_2 = \sigma_1 + \alpha$ we have $\mathbf{L}^{\sigma_2} e^{\alpha s}|\omega| \geq 0$.

This comes from a straightforward computation in a local system of coordinates. Note that the parts *b*) in the preceeding lemma are not completely clear in a point where g (or ω) vanish : then, we may either replace the function $|g|$ by the function $\sqrt{g^2 + \varepsilon^2} - \varepsilon$ (the same for $|\omega|$, or understand this statement in the distribution sense (then, the assertion is that the result is a positive measure).

As a corollary, we get the following

Proposition .—

1) For every pair σ_1 and σ_2 of positive reals such that $\sigma_2^2 \leq \sigma_1^2$, and for $\alpha = \sigma_2 - 2\sigma_1 + \sqrt{2\sigma_1^2 - \sigma_2^2}$, we have

$$|\mathbf{Q}_t^{\sigma_1} f|^2 \leq e^{-\alpha s}\mathbf{Q}_t^{\sigma_2}|f|^2.$$

2) We have a similar statement for forms : for $\sigma_2^2 \leq 2(\sigma_1^2 + \rho)$, and $\alpha = \sigma_2 - 2\sigma_1 + \sqrt{2(\sigma_1^2 + \rho) - \sigma_2^2}$, we have

$$|\vec{\mathbf{Q}}_t^{\sigma_1}\omega|^2 \leq e^{-\alpha s}\mathbf{Q}_t^{\sigma_2}|\omega|^2.$$

We will use these computation in the next part, when dealing with the Riesz transforms in $\mathbf{L}^p$.

3— The RIESZ transforms in $\mathbf{L}^p$.

Comparison between the operators $\mathbf{C}^\sigma$

We are going to give majorizations of the form $\|df\|_p \leq C(p)\|(\mathbf{C}^\sigma - \gamma\mathbf{I})f\|_p$, with different values of the parameters σ and γ depending on the values of the minorant ρ on the RICCI curvature of $\mathbf{L}$. In order to be able to compare these results, our first task is to compare the operators $\mathbf{C}^\sigma$ in $\mathbf{L}^p$. This comparison relies on the following fact : if a function $\varphi(x)$ is the LAPLACE transform of a bounded measure μ on $\mathcal{R}_+$ with total mass $|\mu|$, then, for every p in $[0, \infty]$, the operator $\varphi(-\mathbf{L})$ is bounded on the space $\mathbf{L}^p$ with norm $|\mu|$. To see that, it is enough to remark that

$$\varphi(-\mathbf{L}) = \int_0^\infty \mathbf{P}_t \, d\mu(t),$$

and that each operator $\mathbf{P}_t$ is a contraction of $\mathbf{L}^p$. We have then the following lemma :

Lemma .—

1) If $0 < \sigma_1 < \sigma_2$, the function $\dfrac{\sqrt{\sigma_1^2 + x} - \sigma_1}{\sqrt{\sigma_2^2 + x} - \sigma_2} - 1$ is the LAPLACE transform of a positive measure of mass $\dfrac{\sigma_2}{\sigma_1} - 1$.

2) For every pair $0 \leq \sigma_1 \leq \sigma_2$ and for every $\alpha > 0$, the function $\dfrac{\sqrt{\sigma_1^2 + x} - \sigma_1}{\sqrt{\sigma_2^2 + x} - \sigma_2 + \alpha}$ is a LAPLACE transform of a bounded measure with mass less than $\frac{1}{\alpha}[\sigma_1 + \sigma_2 + \alpha + |\sigma_2 - \alpha - \sigma_1|]$.

Proof. First we look at 1) : we have only to proove that the measure is positive because then the total mass is given by the value of the LAPLACE transform in 0. Then, by a scaling argument, we reduce the problem to the case where $\sigma_2 = 1$, and we write σ instead of σ_1.

Set $u(x) = \sqrt{1+x} - 1$, so that the function under consideration is $(\sqrt{\sigma^2 + 2u + u^2} - \sigma - u)/u$. Since the function $\exp(-tu)$ is the LAPLACE transform of the probability $h(t, s, \sigma)ds$ introduced in the previous chapter, it is enough to show that the function $(\sqrt{\sigma^2 + 2x + x^2} - \sigma - x)/x$ is the LAPLACE transform of a positive bounded measure $\nu(ds)$, because then the measure we are looking for is $\int_t h(t, s, \sigma)\nu(dt)ds$.

Now, consider the function

$$\sqrt{\sigma^2 + 2u + u^2} = (u+1)\sqrt{1 - \frac{1-\sigma^2}{(u+1)^2}}.$$

For $|y| < 1$, we may write $\sqrt{1-y^2} = 1 - \sum_1^\infty \alpha_n y^{2n}$, the coefficient α_n being positive with $\sum \alpha_n = 1$. The function $\sqrt{1-\sigma^2}(1+x)^{-1}$ being the LAPLACE transform of the measure $\rho = \sqrt{1-\sigma^2}\exp(-t)dt$, the function $(1+x) - \sqrt{\sigma^2 + 2x + x^2}$ is the LAPLACE transform of the measure $\rho_1 = \sum_1^\infty \alpha_n \rho^{*(2n-1)}$, which is positive and has mass $(1-\sigma)$. Then, the function $\left((\sigma+x) - \sqrt{\sigma^2+2x+x^2}\right)/x$ is the LAPLACE transform of the measure $\left((1-\sigma)\delta_0 - \rho_1\right) * 1 = (1-\sigma)1_{\{s\geq 0\}}ds - \rho_1 * 1$. Since the measure ρ_1 has total mass $(1-\sigma)$, the measure $\rho_1 * 1$ is $h(t)dt$, with $h(t) = \int_0^t \rho_1(ds) \leq 1-\sigma$. ∎

We now proove the part 2) : we first remark that for every $\sigma \geq 0$ and every $\alpha > 0$, the function $\alpha/(\alpha + \sqrt{\sigma^2 - x} - \sigma)$ is the LAPLACE transform of the probability measure $\int_t \exp(-\alpha t)h(t,s,\sigma)dtds$. Also, the function $\sqrt{1+x} - \sqrt{x}$ is the LAPLACE transform of the probability measure $(2\pi)^{-1}t^{-3/2}(1-e^{-t})dt$, so the function $\frac{1}{\sigma}(\sqrt{\sigma^2+x} - \sqrt{x})$ is also the LAPLACE transform of a probability measure. Now, we write

$$\frac{\alpha}{\sigma_1}\frac{\sqrt{\sigma_1^2+x}-\sigma_1}{\sqrt{\sigma_2^2+x}-\sigma_2+\alpha} = \frac{\sqrt{\sigma_1^2+x}-\sqrt{x}}{\sqrt{\sigma_2^2+x}-\sigma_2+\alpha} +$$
$$\frac{\alpha}{\sigma_1}\left(\frac{\sqrt{x}-\sqrt{\sigma_2^2+x}}{\sqrt{\sigma_2^2+x}-\sigma_2+\alpha} + 1 + \frac{\sigma_2-\alpha-\sigma_1}{\sqrt{\sigma_2^2+x}-\sigma_2+\alpha}\right).$$

This is the LAPLACE transform of a measure whose total mass is less than $1 + \sigma_2/\sigma_1 + \alpha/\sigma_1 + |\sigma_2 - \alpha - \sigma_1|/\sigma_1$. ∎

Notice that, unlike the previous estimate, this bound is not very sharp : when $\sigma_1 = \sigma_2$, we get $2(\alpha+\sigma_1)/\alpha$ instead of 2, which is the right value.

Corollary .—For $0 \leq \sigma_1 \leq \sigma_2$, $\alpha > 0$ and every $p \in [1,\infty]$, a function f is the domain of $\mathbf{C}^{\sigma_1}$ in $\mathbf{L}^p$ if and only if it is in the domain of $\mathbf{C}^{\sigma_2}$ in $\mathbf{L}^p$, and we have

$$\frac{\sigma_1}{\sigma_2-\sigma_1}\|(\mathbf{C}^{\sigma_1} - \mathbf{C}^{\sigma_2})f\|_p \leq$$
$$\frac{\sigma_1+\sigma_2+\alpha+|\sigma_1-\sigma_2-\alpha|}{\alpha}\|(\mathbf{C}^{\sigma_1} - \alpha\mathbf{I})f\|_p.$$

In particular, we have $\|\mathbf{C}^{\sigma_1} f\|_p \leq (\sigma_2/\sigma_1)\|\mathbf{C}^{\sigma_2} f\|_p$ and also

$$(1/3)\big[\sigma\|f\|_p + \|\mathbf{C}^0 f\|_p\big] \leq \|(\mathbf{C}^\sigma - \sigma)f\|_p \leq 2\big[\sigma\|f\|_p + \|\mathbf{C}^0 f\|_p\big].$$

Remark.—

As we may see at once from the proof of the previous lemma, we also have a norm equivalence

$$\gamma\|f\|_p + \frac{1}{2}\|\mathbf{C}^\sigma f\|_p \leq \|(\mathbf{C}^\sigma - \gamma\mathbf{I})f\|_p \leq \gamma\|f\|_p + \|\mathbf{C}^\sigma f\|_p.$$

From these facts, we see that, up to a constant, we may always compare the norms $\|(\mathbf{C}^{\sigma_1} - \alpha_1)f\|_p$ and the norms $\|(\mathbf{C}^{\sigma_2} - \alpha_2)f\|_p$, for different values of the coefficients, and also that, if we know an estimate on the norm $\|\mathbf{C}^\sigma f\|_p$, we also know an estimate on the norms $\|\mathbf{C}^{\sigma_1} f\|_p$ for every σ_1 in the interval $(0, \sigma]$.

Majorizations $\|df\|_p \leq C(p)\|(\mathbf{C}^\sigma - \gamma)f\|_p$).

Everything relies on the following LITTLEWOOD-PALEY inequalities. First, let us introduce the following functions on $[0, \infty)$

$$\begin{cases} V^\sigma(t) &= \frac{1-e^{-2\sigma t}}{2\sigma} \quad \text{for } \sigma > 0; \\ V^\sigma(t) &= s \quad \text{for } \sigma = 0. \end{cases}$$

If $f(t)$ is a function on $\mathcal{R}_+$ with 2 continuous derivatives wich tends to 0 at infinity together with $V^\sigma(t)f'(t)$, then we have

$$f(0) = \int_0^\infty \Big(\frac{\partial^2}{\partial t^2} - 2\sigma\frac{\partial}{\partial t}\Big)f(t)\, V^\sigma(t)\, dt.$$

This explains the role of the function V^σ in the following proposition. We use the same notation $\mathbf{L}^\sigma$ as in the previous chapter.

Proposition .—(LITTLEWOOD-PALEY inequalities)

a) Let f be a bounded positive smooth function on $\mathbf{E} \times \mathcal{R}_+$ such that $\mathbf{L}^\sigma(f^2) \geq 0$. Then, for each $p \in [2, \infty)$, we have

$$\|\Big[\int_0^\infty \{\mathbf{Q}_t^\sigma \mathbf{L}^\sigma(f^2)\}(\cdot, t) V^\sigma(t)\, dt\Big]^{1/2}\|_p \leq C(p)\|f(\cdot, 0)\|_p.$$

Here, the constant $C(p)$ is universal and depends only on p.

b) Assume that $\mathbf{L}^\sigma f \geq 0$ and that

$$\langle \int_0^\infty (|\mathbf{L}^\sigma f^2| + |\mathbf{L} f^2| + |\frac{\partial}{\partial t} f|^2 + |\nabla_x f|^2)\, V^\sigma(t)\, dt\rangle < \infty.$$

Then, for each $p \in (1, 2]$, we have

$$\|\Big[\int_0^\infty \{\mathbf{L}^\sigma(f^2)\}(\cdot, t) V^\sigma(t)\, dt\Big]^{1/2}\|_p \leq C(p)\|f(\cdot, 0)\|_p.$$

Once again, this constant $C(p)$ depends only on p.

We can find the proof of these 2 inequalities in [B4]. They are easy : a) is a direct application of martingales inequalities (very similar to BURKHOLDER inequalities), and the part b) is a direct consequence of the maximal inequality for Markovian semigroups.

We may now give the proof of the inequalities for the RIESZ transforms in the case of a RICCI curvature of $\mathbf{L}$ bounded from below. As before, we denote by ρ a lower bound on the RICCI curvature. Since we are interested in different kind of results (estimates on the operators $\mathbf{C}^\sigma$ for different values of σ and for different values of ρ), we will give a rather complicated proof, involving some extra parameters that we will adjust later according to our needs.

We start with a $\sigma \geq 0$ and a $\gamma > 0$ and write, for a function f in C_c^∞

$$\begin{aligned}
f &= \int_0^\infty \frac{\partial^2}{\partial t^2}(e^{-\gamma t}\mathbf{Q}_t^\sigma f) t\, dt = \int_0^\infty e^{-\gamma t}(\frac{\partial}{\partial t} - \gamma \mathbf{I})(\frac{\partial}{\partial t} - \gamma \mathbf{I})(\mathbf{Q}_t^\sigma f) t\, dt \\
&= \int_0^\infty e^{-\gamma t}(\mathbf{C}^\sigma - \gamma \mathbf{I})^2(\mathbf{Q}_t^\sigma f) t\, dt = \int_0^\infty e^{-\gamma t}\mathbf{Q}_t^\sigma(\mathbf{C}^\sigma - \gamma \mathbf{I})^2(f) t\, dt \\
&= 4\int_0^\infty e^{-2\gamma t}\mathbf{Q}_{2t}^\sigma(\mathbf{C}^\sigma - \gamma \mathbf{I})^2(f) t\, dt \\
&= 4\int_0^\infty e^{-2\gamma t}(\mathbf{C}^\sigma - \gamma \mathbf{I})\mathbf{Q}_t^\sigma \circ \mathbf{Q}_t^\sigma(\mathbf{C}^\sigma - \gamma \mathbf{I})(f) t\, dt.
\end{aligned}$$

In order for this formula to be valid, we need $\gamma > 0$, but later, we will allow γ to be 0 by passing to the limit when this is possible. Then, we have

$$df = 4\int_0^\infty e^{-2\gamma t} d\{(\mathbf{C}^\sigma - \gamma\mathbf{I})\mathbf{Q}_t^\sigma \circ \mathbf{Q}_t^\sigma(\mathbf{C}^\sigma - \gamma\mathbf{I})(f)\} t\, dt$$
$$= 4\int_0^\infty e^{-2\gamma t}(\vec{\mathbf{C}}^\sigma - \gamma\mathbf{I})\vec{\mathbf{Q}}_t^\sigma d\{\mathbf{Q}_t^\sigma(\mathbf{C}^\sigma - \gamma\mathbf{I})(f)\} t\, dt.$$

Take now a 1-form ω in C_c^∞ and consider the scalar product $\langle df, \omega\rangle$. We have

$$\langle df, \omega\rangle = 4\int_0^\infty e^{-2\gamma t}\langle(\vec{\mathbf{C}}^\sigma - \gamma\mathbf{I})\vec{\mathbf{Q}}_t^\sigma d\{\mathbf{Q}_t^\sigma(\mathbf{C}^\sigma - \gamma\mathbf{I})(f)\}, \omega\rangle t\, dt$$
$$= 4\int_0^\infty e^{-2\gamma t}\langle d\{\mathbf{Q}_t^\sigma(\mathbf{C}^\sigma - \gamma\mathbf{I})(f)\}, (\vec{\mathbf{C}}^\sigma - \gamma\mathbf{I})\vec{\mathbf{Q}}_t^\sigma \omega\rangle t\, dt$$
$$= 4\langle\int_0^\infty e^{-2\gamma t} d\{\mathbf{Q}_t^\sigma(\mathbf{C}^\sigma - \gamma\mathbf{I})(f)\} \cdot (\vec{\mathbf{C}}^\sigma - \gamma\mathbf{I})\vec{\mathbf{Q}}_t^\sigma \omega\rangle t\, dt\rangle.$$

We then choose a parameter μ that we will fix later and we set $\mu + \nu = 2\gamma$. We write

$$A = \int_0^\infty e^{-2\mu t}|d\mathbf{Q}_t^\sigma(\mathbf{C}^\sigma - \gamma\mathbf{I})f|^2\, t\, dt, \quad \text{and}$$

$$B = \int_0^\infty e^{-2\nu t}|\vec{\mathbf{Q}}_t^\sigma(\vec{\mathbf{C}}^\sigma - \gamma\mathbf{I})\omega|^2\, t\, dt = \int_0^\infty e^{-2\nu t}|(\frac{\partial}{\partial t} - \gamma\mathbf{I})\vec{\mathbf{Q}}_t^\sigma \omega|^2\, t\, dt,$$

such that we have $\langle df, \omega\rangle \leq 4\langle A^{1/2}B^{1/2}\rangle \leq 4\|A^{1/2}\|_p\|B^{1/2}\|_q$, where q is the conjuguate of p.

We suppose first that $1 < p \leq 2$ and we will proove that, provided that our choice of σ, γ, and μ is accurate, we have $\|A^{1/2}\|_p \leq C(p)\|(\mathbf{C}^\sigma - \gamma\mathbf{I})f\|_p$ and that $\|B^{1/2}\|_q \leq C(q)\|\omega\|_q$, with constants $C(p)$ and $C(q)$ independant of the choice of f and ω. So we end up with an inequality

$$\langle df, \omega\rangle \leq C\|(\mathbf{C}^\sigma - \gamma\mathbf{I})f\|_p\|\omega\|_q,$$

wich prooves that $\|df\|_p \leq C\|(\mathbf{C}^\sigma - \gamma\mathbf{I})f\|_p$.

Majorization of $\|A^{1/2}\|_p$.

We choose a parameter α such that $\mu+\alpha>0$ and $-\sigma\le\alpha\le 0$. We set $\hat{g}=\mathbf{Q}_t^\sigma(\mathbf{C}^\sigma f)$. By the lemma of the last chapter, we have $\mathbf{L}^{\sigma+\alpha}(e^{\alpha t}|\hat{g}|)\ge 0$ and $\mathbf{L}^{\sigma+\alpha}(e^{2\alpha t}|\hat{g}|^2)\ge 2e^{2\alpha t}|d\hat{g}|^2$ so that

$$A\le\frac{1}{2}\int_0^\infty e^{-2\alpha(\mu+\alpha)t}\mathbf{L}^{\sigma+\alpha}(e^{2\alpha t}|\hat{g}|^2)\,t\,dt.$$

Now, notice that we have

$$\frac{s\,e^{-2(\mu+\alpha)s}}{V^{\sigma+\alpha}(s)}\le C[1+\frac{\sigma+\alpha}{\mu+\alpha}].$$

where the constant c is universal. Then we write

$$A\le C'[1+\frac{\sigma+\alpha}{\mu+\alpha}]\int_0^\infty\mathbf{L}^{\sigma+\alpha}(e^{2\alpha t}|\hat{g}|^2)\,V^{\sigma+\alpha}(t)\,dt.$$

It remains to use the LITTLEWOOD-PALEY inequality to get

$$\|A^{1/2}\|_p\le C(p)[1+\frac{\sigma+\alpha}{\mu+\alpha}]^{1/2}\|(\mathbf{C}^\sigma-\gamma\mathbf{I})f\|_p,$$

where the constant $C(p)$ depends only on p.

Majorization of $\|B^{1/2}\|_q$.

We first choose a σ_1 such that $\sigma_1^2\le 2(\sigma^2+\rho)$, wich restricts our choice of σ to those such that $(\sigma^2+\rho)\ge 0$. We set $\hat{\omega}$ for $\vec{\mathbf{Q}}_t^{\sigma_\omega}$. We then use the majorization given in the last chapter and write

$$\begin{aligned}B&=4\int_0^\infty e^{-4\nu t}|\vec{\mathbf{Q}}_{2t}^\sigma(\vec{\mathbf{C}}^\sigma-\gamma\mathbf{I})\omega|^2t\,dt\\&\le 4\int_0^\infty e^{\beta-4\nu t}\mathbf{Q}_t^{\sigma_1}|\vec{\mathbf{Q}}_t^\sigma(\vec{\mathbf{C}}^\sigma-\gamma\mathbf{I})\omega|^2t\,dt\\&=4\int_0^\infty e^{\beta-4\nu t}\mathbf{Q}_t^{\sigma_1}|(\frac{\partial}{\partial t}-\gamma\mathbf{I})\vec{\mathbf{Q}}_t^\sigma\omega|^2t\,dt,\end{aligned}$$

where β is $2\sigma-\sigma_1-\sqrt{2(s^2+\rho)-\sigma_1^2}$. Then we set $\lambda=\sigma_1-\gamma-\sigma$ and we choose σ_1 such that

$$\lambda^2-2\lambda(\sigma_1-2\sigma)+2((\sigma_1-\sigma)^2\le 2\rho\quad\text{and}\quad\beta-4\nu-\lambda<0.$$

Always thanks to the lemma of the previous chapter, we have

$$0 \leq 2|(\frac{\partial}{\partial t} - (\gamma + 2\sigma)\mathbf{I})\hat{\omega}|^2 \leq \mathbf{L}^{\sigma_1}(e^{\lambda t}|\hat{\omega}|^2).$$

From this, we get

$$\begin{aligned} B &\leq 2\int_0^\infty e^{(\beta-4\nu-\lambda)t}\mathbf{Q}_t^{\sigma_1}\mathbf{L}^{\sigma_1}(e^{\lambda s}|\hat{\omega}|^2)s\,ds \\ &\leq C[1+\frac{\sigma_1}{\lambda+4\nu-\beta}]\int_0^\infty \mathbf{Q}_t^{\sigma_1}\mathbf{L}^{\sigma_1}(e^{\lambda s}|\hat{\omega}|^2)V^{\sigma_1}(t)\,dt. \end{aligned}$$

Now, from the LITTLEWOOD-PALEYinequality, it follows that

$$\|B^{1/2}\|_q \leq C(q)[1+\frac{\sigma_1}{\lambda+4\nu-\beta}]^{1/2}\|\omega\|_q,$$

where the constant $C(q)$ depends only on q.

It remains to choose our constants σ, γ, λ, μ, α, σ_1 according to the situation.

1) Case $\rho > 0$: for any $\sigma \geq 0$, we may choose $\gamma = 0$, $\lambda = 0$ and $\sigma_1 = \sqrt{\sigma^2 + 2\rho}$. After optimizing in α and μ, we get a majorization of the form

$$\|df\|_p \leq C(p)(1+\frac{\sigma^2}{\rho})\|\mathbf{C}^\sigma f\|_p,$$

with a universal constant $C(p)$. But in fact, we can get a better estimate if we use the majorization $\|\mathbf{C}^1 f\|_p \leq \sigma\|\mathbf{C}^\sigma f\|_p$, for $\sigma > 1$. Then, once we have got the result for $0 \leq \sigma \leq 1$, we pass to the results for large σ, and we get

$$\|df\|_p \leq C(p)(1+\frac{\sigma}{\sqrt{\rho}})\|\mathbf{C}^\sigma f\|_p.$$

2) Case $\rho = 0$: then we may choose all the parameters to be 0, and we get

$$\|df\|_p \leq C(p)\|\mathbf{C}^0 f\|_p.$$

Then, we may get results for other values of σ using the comparison results given above.

3) Case $\rho < 0$: then, we must choose $\sigma^2 + \rho \geq 0$. We also choose $\gamma = \sigma$ and $\sigma_2 = 0$. This is the case wich we dealed with in [B2]. All the results we get are equivalent to

$$\|df\|_p \leq C(p)\|(\mathbf{C}^{\sigma_0} - \sigma_0 \mathbf{I})f\|_p, \quad \text{with } \sigma_0 = (-\rho)^{1/2},$$

or equivalently

$$\|df\|_p \leq C(p)(\|\mathbf{C}^0 f\|_p + (-\rho)^{1/2}\|f\|_p).$$

It remains to deal with the case $p > 2$. The majorizations are very similar and are left to the reader : we would get exactly the same results as before, for the different values of ρ.

Majorizations $\|\mathbf{C}^\sigma f\|_p \leq C(p)\|df\|_p$.

Once we get the majorizations of $\|df\|_p$, for a given p, a reverse majorization may be obtained for the conjugate exponent q of p. We will show this on 2 examples :

Example 1— Suppose that we know that an inequality $\|df\|_p \leq C\|\mathbf{C}^\sigma f\|_p$ is valid for every f in C_c^∞. Let q be the conjugate exponent of p. Then we have, for every f in C_c^∞ without invariant part

$$\|(\mathbf{C}^\sigma - 2\sigma\mathbf{I})f\|_q \leq C\|df\|_q,$$

with the same constant C. Here, the meaning of "without invariant part" is that, if the measure μ is finite, the function f must have mean 0. (There is no restriction when the measure is infinite.) Obviously, this restriction is nessecary since this result would be wrong for constant functions.

To see this, we use the estimate

$$\|(\mathbf{C}^\sigma - 2\sigma\mathbf{I})f\|_q = \sup_{\|g\|_p \leq 1} \langle(\mathbf{C}^\sigma - 2\sigma\mathbf{I})f, g\rangle.$$

Here, we may restrict ourself to the functions g wich are in $\mathbf{L}_0^p$ (i.e. without invariant part). Moreover, since the operator $\mathbf{C}^\sigma$ is the generator of a Markovian semigroup, it maps C_c^∞ into a dense subspace of $\mathbf{L}_0^p$. Therefore, we may restrict our attention to the functions g of the form $\mathbf{C}^\sigma h$, with in C_c^∞. Then, we write

$$\langle(\mathbf{C}^\sigma - 2\sigma\mathbf{I})f, g\rangle = \langle(\mathbf{C}^\sigma - 2\sigma\mathbf{I})f, \mathbf{C}^\sigma h\rangle = \langle\mathbf{C}^\sigma \text{o}(\mathbf{C}^\sigma - 2\sigma\mathbf{I})f, h\rangle.$$

But the operator $\mathbf{C}^\sigma \circ (\mathbf{C}^\sigma - 2\sigma\mathbf{I})$ is equal to -$\mathbf{L}$, and we may apply the integration by parts formula to get

$$\langle(\mathbf{C}^\sigma - 2\sigma\mathbf{I})f, g\rangle = \langle df.dh\rangle \le \|df\|_q\|dh\|_p \le C\|df\|_p\|\mathbf{C}^\sigma h\|_q \le C\|df\|_q.$$

Now, by using the comparizon results, we also get

$$\frac{1}{4}(\sigma\|f\|_q + \|\mathbf{C}^\sigma f\|_p \le \|(\mathbf{C}^\sigma - 2\sigma\mathbf{I})f\|_p \le C\|df\|_p.$$

So that we see that the majorization $\|df\|_p \le C\|\mathbf{C}^\sigma f\|_p$ leads to an inequality $\sigma\|f\|_q \le C\|df\|_q$. This explains why there are no such inequalities under the unique hypothesis that the RICCI curvature of $\mathbf{L}$ is positive.

Example 2— In the same way, if we have a majorization $\|df\|_p \le C\|(\mathbf{C}^\sigma - \sigma\mathbf{I})f\|_p$, we are led to a reverse inequality

$$\|(\mathbf{C}^\sigma - \sigma\mathbf{I})f\|_q \le C(\|df\|_q + \sigma\|f\|_q).$$

4— The subharmonicity property and the $\mathbf{H}^1$ theory

As we saw in the previous chapter, the $\mathbf{L}^p$ theory of RIESZ transforms rely on the notion of RICCI curvature associated with an elliptic second order differential operator. In order to deal with the $\mathbf{H}^1$ theory, we will introduce another related notion, the **dimension** of such an operator. Remind first that we are dealing with operator wich are elliptic and symmetric, wich means that we may write them in the form $\mathbf{L} = \Delta + \nabla h$, where Δ is the LAPLACE -BELTRAMI operator associated with a Riemannian metric g. Remind also that p denotes the dimension of the manifold.

Definition.—*Let $n \ge p$ be a real number and let ρ be a real function. We will say that the pair (n, ρ) is an admissible pair (dimension,* RICCI*) for* $\mathbf{L}$ *if and only if*

$$\nabla h \otimes \nabla h \le (n-p)[Ric(\mathbf{L}) - \rho g],$$

where this inequality has to be understood in the sense of symmetric tensors (i.e. if we compute their difference in a local system of coordinates, it is represented by a positive matrix.)

Examples.

1) Let $h = 0$, so that $\mathbf{L}$ is the LAPLACE -BELTRAMI operator of the metric g. Then, (n, ρ) is admissible iff $n \geq p$ and $\rho(x)$ is less than the lowest eigenvalue $r(x)$ of $Ric(\Delta)$ at the point x. In this case, there is a best admissible pair (p, r).

2) $\mathbf{L}$ is an operator on the real line (or an interval), written in the canonical form $\frac{\partial^2}{\partial t^2} - a(t)\frac{\partial}{\partial t}$. Then (n, ρ) is admissible iff

$$\frac{\partial}{\partial t}a \geq \rho + \frac{a^2}{n-1}.$$

For example, for the ultraspherical operators of order ν that we considered in the first chapter, we have, in the canonical form

$$\frac{\partial}{\partial t}a = \nu + \frac{a^2}{\nu - 1}.$$

When n is an integer, this reflects the fact that they are the projections of the LAPLACE -BELTRAMI operator of the spheres of radius 1 and dimension n, which have RICCI curvature n.

Remark.—

If we remind the definition of the operator Γ_2 introduced in the 2nd chapter, then the lower eigenvalue r of the RICCI curvature was characterized by the inequality $\Gamma_2(f, f) \geq r\Gamma(f, f)$. There is the same intrinsic definition for the dimension : a pair (n, ρ) is admissible for $\mathbf{L}$ iff, for every f in $\mathcal{C}_c^\infty$, we have

$$\Gamma_2(f, f) \geq \frac{1}{n}(\mathbf{L}f)^2 + \rho\Gamma(f, f).$$

The subharmonicity property is related to the notion of dimension in the following way :

Lemma .—Suppose that (n, ρ) is admissible for $\mathbf{L}$ and that f is a function wich satisfy $\mathbf{L}f = 0$ in an open set Ω. Then we have (*cf* [B3])

$$\forall p \in [\frac{n-1}{n-2}, 1], \quad (\mathbf{L} - p\rho\mathbf{I})|\nabla f|^p \geq 0 \text{ in } \Omega.$$

We will not proove this property, which follows from a computation in a local system of coordinates. In general, we do not apply this theorem on $\mathbf{E}$ itself, but on $\mathbf{E} \times \mathcal{R}_+$, with the operators $(\mathbf{L}^\sigma)$ and the functions of the form $\hat{f}(x,t) = \mathbf{Q}_t^\sigma f(x)$ wich are solutions of $(\mathbf{L}^\sigma)\hat{f} = 0$. Such functions are also solutions, for every β, of the equation $e^{2\beta t}(\mathbf{L}^\sigma) f = 0$. This explains why we have to show how to pass from the pairs (n, ρ) admissible for $\mathbf{L}$ to the pairs admissible for $e^{2\beta t}(\mathbf{L}^\sigma)$. This comes from the following

Proposition .—Suppose that the constant pair (n, ρ) is admissible for $\mathbf{L}$. Then, when $4\sigma^2 \geq (n-1)\rho$, and $\beta = \dfrac{2\sigma}{n-1}$, the pair $(n+1, e^{2\beta t}[\rho - \dfrac{4\sigma^2}{n-1}])$ is admissible for $e^{2\beta t}\mathbf{L}^\sigma$.

As an application, choose a σ such that $4\sigma^2 \geq (n-1)\rho$, and take the corresponding value of β. For a given function f in $\mathcal{C}_c^\infty$, we set $\hat{f} = \mathbf{Q}_t^\sigma f(x)$. Then we have

$$(\mathbf{L}^\sigma - \frac{n}{n-1}(\rho - \frac{4\sigma^2}{n-1})\mathbf{I})\{e^{2\beta t}(|\nabla \hat{f}|^2 + |\frac{\partial}{\partial t}\hat{f}|^2)\}^{(n-1)/(2n)} \geq 0.$$

Provided that we also choose $(n-2)\sigma^2 + (n-1)\rho \geq 0$, we may set $\delta = \sqrt{\dfrac{(n-2)\sigma^2 + (n-1)\rho}{n}}$ and $\gamma = \delta - \dfrac{n-1}{n}\sigma$. Then the last inequality becomes

$$(\frac{\partial^2}{\partial t^2} - 2\delta\frac{\partial}{\partial t} + \mathbf{L})e^{\gamma t}(|\nabla \hat{f}|^2 + |\frac{\partial}{\partial t}\hat{f}|^2)^{(n-1)/(2n)} \geq 0.$$

If we also choose σ such that $\sigma^2 \geq n(n-1)\rho$, then the function $e^{\gamma t}(|\nabla \hat{f}|^2 + |\frac{\partial}{\partial t}\hat{f}|^2)^{(n-1)/(2n)}$ is bounded and we may conclude that

$$(|\nabla \mathbf{Q}_t^\sigma f|^2 + |\frac{\partial}{\partial t}\mathbf{Q}_t^\sigma f|^2)^{(n-1)/(2n)} \leq e^{-\gamma t}\mathbf{Q}_t^\delta\{(\mathbf{C}^\sigma f)^2 + |\nabla f|^2\}^{\frac{n-1}{2n}}.$$

We may now use the maximal inequality for the symmetric MARKOV semigroup $\mathbf{Q}_t^\delta$, with the exponent $n/(n-1)$ to get

$$\|\sup_t e^{\gamma t}|\mathbf{Q}_t^\sigma \mathbf{C}^\sigma f|\|_1 \leq C(n)(\|\mathbf{C}^\sigma f\|_1 + \|\nabla f\|_1).$$

In this last formula, we may want to replace the function f by the function $f = (\mathbf{C}^\sigma)^{-1}g$, in order to get an assertion about the RIESZ transforms in $\mathbf{L}^1$ as in the classical cases. But there is a difficulty there : the space generated by the functions $\mathbf{C}^\sigma f$ with f in $\mathcal{C}_c^\infty$ is not in general dense in $\mathbf{L}^1$. It's closure is the orthogonal of the closed subspace of $\mathbf{L}^\infty$ formed by bounded $\mathbf{L}$-harmonic functions. (It will be dense in $\mathbf{L}_0^1$, for example as soon as the invariant measure μ is finite.) Let us denote this closed subspace of $\mathbf{L}^1$ by $\mathbf{L}_{00}^1$. Then, on this subspace, the operator $\vec{\mathbf{R}}^\sigma = \nabla(\mathbf{C}^\sigma)^{-1}$ is a densely defined closed operator and we have

$$\| \sup_t e^{\gamma t} |\mathbf{Q}_t^\sigma f| \|_1 \leq C(n)(\|f\|_1 + \|\vec{\mathbf{R}}^\sigma f\|_1).$$

To get reversed results, we will restrict ourself to the case $\rho = 0$: this means that we are now assuming thar the operator $\mathbf{L}$ admits a constant pair $(n, 0)$ as (dimension, curvature). Then, we also restrict ourself to the case $\sigma = 0$, and we omit the 0 in the subsequent notations. In this case, the previous results show that we have a majorization

$$\| |\mathbf{Q}_t f| \|_1 \leq C(n)(\|f\|_1 + \|\vec{\mathbf{R}} f\|_1).$$

We introduce a diffusion process (X_t) on $\mathbf{E}$, with generator $\mathbf{L}$ and initial law μ. The fact that μ may be infinite introduce a few more complications, but nothing serious. We also introduce a independant Brownian motion (B_t) on $\mathcal{R}_+$, starting from a real $a \geq 0$ and we stop it at the first exit time T_0 of the positive half line. On $\mathbf{E} \times \mathcal{R}_+$, we may consider the process $Z_t = (X_{t\wedge T_0}, B_{t\wedge T_0})$ wich has generator $(\frac{\partial^2}{\partial t^2} + \mathbf{L})$.

When f is a given function on $\mathbf{E}$, consider it's harmonic prolongation $\hat{f}(x,t) = \mathbf{Q}_t f(x)$: then the process $M_t^f = \hat{f}(Z_t)$ is a martingale. We denote by $\|f\|_{\mathbf{H}^{1a}}$ the norm $E[\sup_t |M_t^f|]$. The MARKOV property shows at once that this is an increasing function of a, and we set

$$\|f\|_{\mathbf{H}^1} = \sup_a \|f\|_{\mathbf{H}^{1a}}.$$

Now , for a function f in the space $\mathbf{L}_{00}^1$, we have the following

Theorem .—There exits a universal constant c, and a constant $C(n)$ depending only on the dimension n, such that, for every operator $\mathbf{L}$ with finite dimension n and positive RICCI curvature, one has

$$c(\|f\|_1 + \|\vec{\mathbf{R}}f\|_1) \leq \|f\|^1_{\mathbf{H}} \leq C(n)(\|f\|_1 + \|\vec{\mathbf{R}}f\|_1).$$

We will not give a proof of this result, which may be found in [B3]. Let us just mention that the proof of the second inequality is very similar to the proof we gave for the disk in the first part, and that the the proof of the second inequality is closed to the proof of the $\mathbf{L}^p$ results of the third part.

There are still a lot of open problems here. First, what kind of $\mathbf{H}^1$ equivalence may be true when the RICCI curvature is just bounded from below (things are much more complicated in the $\mathbf{H}^1$ case when we deal with the semigroups $\mathbf{Q}^\sigma_t$ instead of the semigroups $\mathbf{Q}^0_t$). The second problem is if there are general equivalences between the norms $\| \sup_t |\mathbf{Q}^0_t f| \|_1$ and the probalistic $\mathbf{H}^1$ norm, as there is in the classical case. There is always a minoration $\| \sup_t |\mathbf{Q}^0_t f| \|_1 \leq \|f\|^1_{\mathbf{H}}$, which is very easy and universal, but the reverse inequality is not known, except in particular cases like $\mathcal{R}^n$ or the spheres.

References.—

[B1] BAKRY (Dominique)— Etude probabiliste des transformées de RIESZ et de l'espace $\mathbf{H}^1$ sur les sphères—**Séminaire de probabilités XVIII** , Lecture Notes in Math. n^o**1059**, Springer, 1983, p. 197-218 .

[B2] BAKRY (Dominique)— Transformations de RIESZ pour les semigroupes symétriques —**Séminaire de Probabilités XIX** , Lecture Notes in Math. n^o**1123**, Springer, 1985, p. 130-174 .

[B3] BAKRY (Dominique)— La propriété de sous-harmonicité des diffusions dans les variétés—**Séminaire de Probabilités XXII**, Lecture Notes in Math., **to appear**.

[B4] BAKRY (Dominique)— Etude des transformations de RIESZ dans les variétés riemanniennes à courbure de RICCI minorée —**Séminaire de Probabilités XXI** , Lecture Notes in Math. *n°***1247**, Springer, 1987, p. 137-172 .

[B5] BAKRY (Dominique) — Un critère de non explosion pour certaines diffusions sur une variété riemannienne complète —Comptes Rendus Acad. Sc., t. **303**, série 1, *n°*1, 1986, p. 23-27 .

[CW] COIFMAN (Ronald R.) et WEISS (Guido)— **Analyse harmonique non commutative sur certains espaces homogènes**—Lecture notes in Math.*n°***242**, Springer,1971.

[L] LOHOUE (Noel)— Comparaison des champs de vecteurs et des puissances du laplacien sur une variété riemannienne à courbure non positive, J. Funct. Anal., vol.**61**, 1985, p. 164-205 .

[M1] MEYER (Paul-André)— Démonstration probabiliste de certaines inégalités de LITTLEWOOD - PALEY —**Séminaire de Probabilités X** , Lecture Notes in Math. *n°* **511**, Springer, 1976, p. 125-183 .

[M2] MEYER (Paul-André)— Transformations de RIESZ pour les lois gaussiennes—**Séminaire de Probabilités XVIII** , Lecture Notes in Math. *n°* **1059**, Springer, 1983, p. 179-193 .

[M3] MEYER (Paul-André)— Notes sur les processus d'ORNSTEIN-UHLENBECK—**Séminaire de Probabilités XVI** , Lecture Notes in Math. *n°* **920**, Springer, 1982, p. 95-132 .

[*M*4] MEYER (Paul-André)— Le dual de $\mathbf{H}^1(\mathbf{R}^\nu)$: démonstrations probabilistes. —**Séminaire de Probabilités XI**, Lecture Notes in Math. *n°* **581**, Springer, 1977, p. 132-195 .

[*RW*] RICCI (Fulvio) ; WEISS (Guido)— A characterization of $\mathbf{H}^1(\mathbf{S}^n)$, **Proc. Symp. Pure Math.**, *AMS* 1979, p. 35- .

[*S*] STEIN (Elias M.) — **Singular integrals and differentiability properties of functions**—Princeton, 1970.

[*Str*] STRICHARTZ (R.)— Analysis of the Laplacian of the complete Riemannian manifold, J.of Functionnal Analysis, vol.**52** , 1983, p. 48-79 .

[*SW*] STEIN (Elias M.) et WEISS (Guido)— On the theory of harmonic functions of several variables, Acta Mathematica, vol.**103**, 1960, p. 25-62 .

Dominique Bakry
IRMA
7, rue René Descartes
67084 STRASBOURG cedex
FRANCE

THE OPTIONAL STOCHASTIC INTEGRAL

By

JAMES K. BROOKS and DAVID NEAL

Introduction and Notation.

In this paper we shall study the optional (or compensated) stochastic integral $H\dot{_c}X$. The two main problems connected with this integral will be considered. First, we wish to express $H\dot{_c}X$ in terms of an ordinary predictable stochastic integral $H'\cdot X$, where H' is a suitable predictable process associated with the optional process H. An attempt in this direction was first undertaken by Yor [8]; however, even for bounded, scalar H, the problem remained open. We shall show in this case that $H\dot{_c}X - H'\cdot X$ exists as a certain limit in M^2, the space of cadlag (Hilbert-valued) square integrable martingales, (cf. §3). Secondly, we shall develop $H\dot{_c}X$ for processes H and X which take their values in a separable Hilbert space. These integrals, in turn, will allow us in a later paper to develop $H\dot{_c}X$ for certain nuclear-valued processes. Full details of the proofs of the theorems presented here will appear elsewhere.

Following the notation of Déllacherie and Meyer [2], we shall work with a probability space (Ω,F,P) having a filtration (F_t) which satisfies the usual conditions. Moreover, we assume $F_{0-} = F_0$ and $F_{\infty-} = F_\infty$. The space E will always denote a real, separable Hilbert space. The symbol $|\cdot|$ will denote either the absolute value, if E is the scalar field, or the norm in E, while $(\cdot,\cdot)$ will denote the inner product in E. We shall make use of the predictable stochastic integral for Hilbert-valued processes, developed first by Kunita [5], and the optional (or compensated) stochastic integral for real processes, presented in Déllacherie and Meyer [2]. Finally, we shall freely use the properties of the square bracket (or quadratic variation) of Hilbert-valued semimartingales (cf. [2] and [6]).

1. The Classical Setting.

In the case when X is an E-valued square integrable martingale and H is a real, optional process, the classical method (cf. [2]) can be used to define the optional integral.

We let $L_0^2(X)$ be the space of real, optional processes H such that

(1) $$\|H\|_X = (E[\int^\infty |H_s|^2 d[X,X]_s])^{1/2} < \infty.$$

We define a linear operator $T: L^2(\Omega,F,P,E) \to R$ by

(2) $$T(Y_\infty) = E[\int^\infty H_s d[X,Y]_s],$$

for every $Y_\infty \in L^2(\Omega,F,P,E)$, where Y is a cadlag version of the square integrable martingale $E(Y_\infty|F_t)$. Using the Kunita-Watanabe inequality and Hölder's inequality, one can show that T is continuous; hence, there exists a unique $W_\infty \varepsilon L^2(\Omega,F,P,E)$ such that

(3) $T(Y_\infty) = E[(W_\infty,Y_\infty)]$

for every $Y_\infty \varepsilon L^2$. We then let $H_{\dot{c}}X$ be a cadlag version of the square integrable martingale $E(W_\infty|F_t)$.

We shall refer to the following equation

(4) $E[((H_{\dot{c}}X)_\infty,Y_\infty)] = E[\int^\infty H_s d[X,Y]_s]$

as the integral characterization of $H_{\dot{c}}X$.

The usual properties of the optional integral (cf. [2]) hold for $H_{\dot{c}}X$ in this case. In particular, $H_{\dot{c}}X$ agrees with the predictable stochastic integral $H\cdot X$ if H is predictable. Moreover, for every $f\varepsilon E$, $(f,H_{\dot{c}}X) = H_{\dot{c}}(f,X)$. Using the theory of H^p spaces and BMO spaces of Hilbert-valued martingales, we can extend the above theory first to local martingales X and then to special semimartingales.

2. The optional integral for Hilbert-valued H and scalar-valued X.

We define $L_0^2(X)$ in this setting as in §1, where X is now a real, square integrable martingale. For $H\varepsilon L_0^2(X)$, we define a linear operator T: $L^2(\Omega,F,P,R) \to E$ by

$$T(Y_\infty) = E[\int^\infty H_s d[X,Y]_s],$$

where Y is a cadlag version of the real square integrable martingale $E(Y_\infty|F_t)$. Again, by the Kunita-Watanabe and Hölder inequalities, one can show that T is continuous. However, we encounter a difficulty in this setting in that the continuity of T is not sufficient to ensure that T has an integral representation. To this end, we need a slight excursion into operator theory.

We let

$$|||T|||_2 = \sup \{ \sum_{i=1}^{n} |a_i T(1_{E_i})| \},$$

where the supremum is taken over all functions $f = \sum_{i=1}^{n} a_i 1_{E_i}$, where (E_i) is a disjoint collection of sets from F and $\|f\|_{L^2} < 1$. Since E is reflexive and thus has the Radon-Nikodým property, it can be shown (cf. [3] and [4]) that the finiteness of $|||T|||_2$ is equivalent to T having an integral representation. In this case, T is a compact operator. Thus, if $|||T|||_2 < \infty$, then there exists a unique $W_\infty \varepsilon L^2(\Omega, F, P, E)$ such that $T(Y_\infty) = E[Y_\infty W_\infty]$, for every $Y_\infty \varepsilon L^2(\Omega, F, P, R)$. Moreover, $\|W_\infty\|_{L^2} = |||T|||_2$. We can then define the stochastic integral $H \underset{c}{\cdot} X$ to be a cadlag version of $E(W_\infty | F_t)$. It is generally impossible though to establish the finiteness of $|||T|||_2$ directly; however, by using an orthonormal expansion of H, we may explicitly produce a density W_∞ for the operator T. We can thus prove the following result:

<u>Theorem 1</u>. <u>Let $H \varepsilon L^2_0 (X)$. There exists a unique E-valued square integrable martingale $H \underset{c}{\cdot} X$, called the optional stochastic integral of H with respect to X, such that for every $Y_\infty \varepsilon L^2(\Omega, F, P, R)$</u>

$$E[H \underset{c}{\cdot} X)_\infty Y_\infty] = E[\int^{\infty} H_s d[X,Y]_s],$$

<u>where Y is a cadlag version of the martingale</u> $E(Y_\infty | F_t)$.

We remark that the existence of $H \cdot_c X$ provides an interesting example of a compact Hilbert-valued operator defined on L^2.

As in the first setting, all of the usual properties of the classical optional stochastic integral remain valid. Also, for every $f \in E$, $(f, H \cdot_c X) = (f, H) \cdot_c X$. This integral can be further extended to the case when H is locally bounded and optional and X is a special semimartingale.

3. *A natural definition of* $H \cdot_c X$ *and the general case.*

In this section, both processes H and X may be either real or Hilbert-valued. Juxtaposition of processes will denote either the product, scalar product, or inner product depending on whether the processes are real and/or E-valued.

To treat these cases, we introduce an alternate method of defining the optional integral which yields a natural, integral-type definition of $H \cdot_c X$ and shows its precise relationship to the predictable stochastic integral.

The first step is to approximate a bounded, E-valued, optional process H by a suitable predictable process H'. This procedure is accomplished by considering the predictable projection $H^{\cdot} = {}^{p}H$ of H, which exists by Brooks-Dinculeanu [1], and then using the separability of E.

Lemma 2. *Let H be a bounded, optional process. There exists a predictable process H' such that* $H_0 = H'_0$ *a.s. and*

H - H' is thin. Moreover, H' may be taken to have the same bound as H.

The next result is crucial for obtaining the desired structure theorem for $H_{\dot{c}}X$.

Theorem 3. Let X be a square integrable martingale, let H be bounded and optional, and let H' be bounded and predictable such that $H_0' = H_0$ a.s. and H - H' is thin. For each n, we set

$$B_t^n = \sum_{s \leq t} (H_s - H_s')\Delta X_s 1_{\{|\Delta X_s| > \frac{1}{n}\}}.$$

Then for each n,

(a) B^n is locally integrable

and

(b) $B^n - (B^n)^p$ is a square integrable martingale.

Moreover, the sequence $(B^n - (B^n)^p)$ is Cauchy in M^2.

Sketch of Proof: Since X is cadlag, the process B_t^n is well-defined.

Since H and H' are bounded by a common constant k, it suffices to show that the process

$$C_t^n = \sum_{s \leq t} |\Delta X_s| 1_{\{|\Delta X_s| > \frac{1}{n}\}}$$

is locally integrable. This result follows immediately from the inequality $C_t^n \leq C_{t-}^n + (\Delta[X,X]_t)^{1/2}$ and the fact that each process on the right hand side is locally integrable. Hence, B_t^n is also locally integrable and $(B^n)^p$ exists.

We next observe that

$$[B^n-(B^n)^p, B^n-(B^n)^p]_\infty \leq 2([B^n,B^n]_\infty + [(B^n)^p,(B^n)^p]_\infty).$$

Then since

$$E[[B^n-(B^n)^p,B^n-(B^n)^p]_\infty]$$

$$< 4E[[B^n,B^n]_\infty]$$

$$= 4E[\sum_{s<\infty} |\Delta B_s^n|^2]$$

$$= 4E[\sum_{s<\infty} |(H_s-H_s')\Delta X_s|^2 1_{\{|\Delta X_s|>\frac{1}{n}\}}]$$

$$< 16k^2 E[[X,X]_\infty]$$

$$< \infty,$$

it follows that the local martingale $B^n - (B^n)^p$ is a square integrable martingale.

Since $\sum_{s<\infty} \Delta[X,X]_s$ belongs to L^1, by an interchange of limits and then the monotone convergence theorem,

$\lim_n \sum_{s<\infty} \Delta[X,X]_s 1_{\{|\Delta X_s|>\frac{1}{n}\}} = \sum_{s<\infty} \Delta[X,X]_s$ both pointwise and in L^1. We set $Y^n = B^n - (B^n)^p$.

Then, for $n > m$,

$$E[[Y^n-Y^m,Y^n-Y^m]_\infty]$$

$$< 4E[\sum_{s<\infty} |\Delta(B_s^n-B_s^m)|^2]$$

$$= 4E[\sum_{s<\infty} |(H_s-H_s')\Delta X_s|^2 1_{\{\frac{1}{m} > |\Delta X_s|>\frac{1}{n}\}}]$$

$$< 16k^2(E[\sum_{s<\infty} \Delta[X,X]_s 1_{\{|\Delta X_s|>\frac{1}{n}\}} - \sum_{s<\infty} \Delta[X,X]_s 1_{\{|\Delta X_s|>\frac{1}{m}\}}]).$$

Thus, (Y^n) is Cauchy in M^2.

<u>Definition</u>. Let H be a bounded, optional process and let X be a square integrable martingale. Let H' be any

bounded, predictable process such that $H'_0 = H_0$ a.s. and $H - H'$ is thin and let B^n_t be defined as in Theorem 3.

We define the optional stochastic integral of H with respect to X to be the square integrable martingale

$$H \underset{c}{\cdot} X = H' \cdot X + Z,$$

where $Z = M^2 - \lim_n (B^n - (B^n)^p)$.

<u>Theorem 4</u>. <u>The optional stochastic integral</u> $H \underset{c}{\cdot} X$ <u>is a well-defined process, independent of the choice of H' in the definition</u>.

We remark that if one or both of H or X is real-valued, then $H \underset{c}{\cdot} X$ defined above agrees with the previous definitions of Sections 1 and 2. Otherwise, we have the following result, whose proof requires several technical lemmas.

<u>Theorem 5</u>. <u>Let H be bounded and optional and let X be a square integrable martingale. If both H and X are E-valued, then the following characterization holds:</u>

$H \underset{c}{\cdot} X$ <u>is the unique real square integrable martingale such that for every real square integrable martingale Y,</u>

$$E[(H \underset{c}{\cdot} X)_\infty Y_\infty] = E[\int^\infty H_s dA^Y_s],$$

<u>where A^Y is the $E^*(=E)$-valued process of integrable variation given by</u>

$$A^Y_t(f) = [Y,(f,X)]_t, \quad f \varepsilon E.$$

<u>Remarks</u>. Using the above characterization, we may obtain all of the usual properties of the optional stochastic integral. If X is continuous and H is optional.

then we see immediately that $H \underset{c}{\cdot} X = H' \cdot X$, where H' is the predictable process appearing in the definition of $H \underset{c}{\cdot} X$. L. Schwartz [7] has considered $H \underset{c}{\cdot} X$ with this condition on X. We see further that if H is predictable, then we may take $H' = H$ and then $H \underset{c}{\cdot} X = H \cdot X$, the predictable stochastic integral.

As before, if H and X are both E-valued, $H \underset{c}{\cdot} X$ can be extended to locally bounded, optional H and special semimartingales X.

BIBLIOGRAPHY

[1] J. K. Brooks and N. Dinculeanu, Projections and regularity of abstract processes, Stochastic Analysis and Applications, 5(1), (1987), pp. 17-25.

[2] C. Déllacherie and P. Meyer, Probabilities and Potential B, North Holland, 1982.

[3] J. Diestel and J. J. Uhl, Jr., Vector Measures, Math. Surveys, No. 15, A.M.S., 1977.

[4] N. Dinculeanu, Vector Measures, Permagon Press, 1967.

[5] H. Kunita, Stochastic integral based on martingales taking values in Hilbert space, Nagoya Math. J., 38 (1970), pp. 41-52.

[6] M. Métivier, Semimartingales, Walter de Gruyter, Berlin, 1982.

[7] L. Schwartz, Semimartingales and their stochastic calculus on manifolds, Les Presses de l'Université de Montréal, 1984.

[8] M. Yor, En cherchant une définition naturelle des intégrales stochastiques optionnelles, Séminaire de Probabilités XIII, Lecture Notes in Math., 721, 1979.

J. K. Brooks and D. Neal
Department of Mathematics
University of Florida
Gainesville, FL 32611

ON BROWNIAN EXCURSIONS IN LIPSCHITZ DOMAINS PART II. LOCAL ASYMPTOTIC DISTRIBUTIONS[1]

KRZYSZTOF BURDZY

ELLEN H. TOBY

RUTH J. WILLIAMS

1. Introduction. In this paper, we continue the study initiated in Burdzy and Williams (1986) of the local properties of Brownian excursions in Lipschitz domains. The focus in part I was on local *path* properties of such excursions. In particular, a necessary and sufficient condition was given for Brownian excursions in a Lipschitz domain to share the local path properties with Brownian excursions in a half-space. This condition holds for $C^{1,\alpha}$-domains ($\alpha > 0$), but there is a C^1-domain for which it fails. Here we consider the *distributions* of a selection of local events for excursions. In particular, we focus on the asymptotics of these distributions as the region of locality shrinks to a point. We show that when a Lipschitz domain is locally approximated by a half-space, the asymptotics for excursions in the two domains are comparable.

Our choice of which local events to examine was influenced by the desire to give a representation for Brownian local time on the boundary of a Lipschitz domain. The definition of local time we use here may seem somewhat unorthodox, however, it is quite natural in the context of exit systems. Specifically, we define the local time to be the continuous additive functional of Brownian motion whose Revuz measure (relative to Lebesgue measure as invariant measure) is equal to the surface area measure on the boundary of the Lipschitz domain (Revuz

[1]Research supported in part by NSF Grants DMS 8419377, DMS 8702620, DMS 8319562 and DMS 8657483.

(1970)). Using the results of parts I and II, for a class of Lipschitz domains that includes the $C^{1,\alpha}$-domains ($\alpha > 0$), we give several representations for Brownian local time in terms of limits of numbers of excursions of a certain "size".

Representation theorems of this type are well known for one-dimensional Brownian motion (see Fristedt and Taylor (1983) or Williams (1979)) and have been obtained for multi-dimensional *reflected* Brownian motion in a C^3-domain (see Hsu (1986), Theorem 6.1). Although we consider unconstrained Brownian motion rather than reflected Brownian motion, we require less regularity of the boundary than Hsu (1986), and it seems that a modification of our approach would yield a relaxation of his assumptions. Another closely related paper is that of Bass (1984) who discussed convergence of *continuous* additive functionals of Brownian motion. His results can be used to give some representation theorems for Brownian local time on a Lipschitz surface. In contrast, the representations we give are in terms of limits of *discontinuous* additive functionals that count the number of excursions of a certain size.

The key to our representations is a link we establish between the local behavior of excursions and the boundary behavior of the Green function. The potential of this connection has not been exhausted here, for further representation theorems could be obtained almost automatically once more about the Green function was known. In fact, some results of this nature have recently been obtained and will be discussed in a forthcoming paper by Bañuelos and Burdzy (1988).

The remainder of this paper is organized as follows. Section 2, introduces notation and reviews some known results. Section 3 presents explicit formulas for some distributions of excursions in a half-space. In Section 4 we show that when a Lipschitz domain is locally approximated by a half-space, the asymptotics of the distributions of a selection of local events for excursions in the two domains are comparable. These results are an improvement of those in Chapter 4 of

Burdzy (1987). Representation theorems for local time are proved in Section 5. Section 6 contains some potential theoretic results which are needed in Section 4. This section relies heavily on the results of Fabes et al. (1986) and does not use probability, hence it is relegated to the end of the paper.

The authors would like to thank Rodrigo Bañuelos and Donald Sarason for their most valuable suggestions.

2. Preliminaries. In this section we establish notation and review some known results that are fundamental to our study. The reader is advised to consult Doob (1984) concerning Brownian motion and potential theory. Fabes et al. (1986) is an excellent reference concerning the boundary behavior of parabolic functions. We use the version of excursion theory presented in Burdzy (1987) which is based on more general results of Maisonneuve (1975).

Since the paper of Fabes et al. (1986) will be referred to repeatedly in Section 6, we adopt its notation with slight modifications.

Let $\mathbf{R}^n$ denote n-dimensional Euclidean space for some $n \geq 2$. The points of $\mathbf{R}^n$ will be denoted as x, y, etc. In describing local properties at different points on the boundary of a domain in $\mathbf{R}^n$, it will be convenient to use orthonormal coordinate systems that vary from point to point. The notation $(x_1, \ldots, x_n)$ will be used to denote the coordinates of a point x in $\mathbf{R}^n$ in some orthonormal coordinate system, where the choice of coordinate system will always be made clear in the context. We will then write $\widetilde{x} = (x_1, \ldots, x_{n-1})$, in particular $\widetilde{0} = (0, \ldots, 0) \in \mathbf{R}^{n-1}$. The complement of a set D in $\mathbf{R}^n$ will be denoted $D^c = \mathbf{R}^n \setminus D$.

A set $D \subset \mathbf{R}^n$ will be called a Lipschitz domain if it is a non-empty domain and there exists $\lambda < \infty$ such that for each $x \in \partial D$ there is a non-empty neighborhood U of x, an orthonormal coordinate system $CS(x)$, and a Lipschitz function $\varphi_x : \mathbf{R}^{n-1} \to \mathbf{R}$ with constant λ, satisfying $D \cap U = \left\{ y \in U : y_n > \varphi_x(\widetilde{y}) \right\}$ in

$CS(x)$. Then for $r > 0$, $x \in \partial D$ and $t \in \mathbf{R}$, let

$$\Psi_r(x,t) = \left\{(y,u) \in D \times \mathbf{R} : |x-y| < r, |t-u| < r^2\right\},$$
$$\Delta_r(x,t) = \overline{\Psi_r(x,t)} \cap (\partial D \times \mathbf{R}),$$

where $\overline{\Psi_r(x,t)}$ denotes the closure of $\Psi_r(x,t)$ in $\mathbf{R}^{n+1}$, and for $x = (x_1, \ldots, x_n)$ in $CS(x)$, let

$$\overline{A}_r(x,t) = \left(\widetilde{x}, x_n + r, t + 2r^2\right),$$
$$\underline{A}_r(x,t) = \left(\widetilde{x}, x_n + r, t - 2r^2\right).$$

For a domain $U \subset \mathbf{R}^n$ and open interval $I \subset \mathbf{R}$, a function $f : U \times I \to \mathbf{R}$ will be called parabolic if the first and second partial derivatives of f in U and first partial derivative of f in I are continuous on $U \times I$, and

$$\sum_{j=1}^{n} \frac{\partial^2}{\partial x_j^2} f(x,t) - \frac{\partial}{\partial t} f(x,t) = 0 \quad \text{for} \quad (x,\ t) \in U \times I.$$

For $(y,u) \in \overline{\Psi_r(x,t)}$, the caloric measure $\mu_{(y,u)}$ on $\partial\Psi_r(x,t)$ is the unique Borel probability measure that does not charge $\{(z,s) \in \partial\Psi_r(x,t) : s = t + r^2\}$ and satisfies

$$f(y,u) = \int_{\partial\Psi_r(x,t)} f(z,s)\mu_{(y,u)}(dz,ds)$$

for every parabolic function f in $\Psi_r(x,t)$ which is continuous in $\overline{\Psi_r(x,t)}$ (Fabes et al. (1986)).

The Green function of a Lipschitz domain $D \subset \mathbf{R}^n$ will be denoted $G_D(\,\cdot\,,\,\cdot\,)$.

Let Ω be the space of paths $\omega : [0,\infty) \to \mathbf{R}^n \cup \{\delta\}$ which are continuous on $[0,R)$ for some $R(\omega) \le \infty$ and such that $\omega(t) = \delta$ for $t \ge R$. Thus, R denotes the lifetime of a path, which may be infinite. Let X be the canonical process i.e., $X_t(\omega) \equiv \omega(t)$. Denote $\mathcal{F} = \sigma\left\{X_t, t \ge 0\right\}, \mathcal{F}_t = \sigma\left\{X_s, s \le t\right\}$. For a stopping time T let $\mathcal{F}_T$ denote the usual σ-field of pre-T-events and let θ_t, $t \ge 0$, be the shift operators on Ω. For a set $A \subset \mathbf{R}^n$ let

$$T_A = T(A) = \inf\{t > 0 : X_t \in A\}$$

and

$$T(A-) = \inf\{t > 0 : \lim_{s\uparrow t} X_s \in A\}.$$

Let P^x denote a measure on $(\Omega, \mathcal{F})$ which makes X the standard n-dimensional Brownian motion starting from x. Analogously, P_D^x will denote the distribution of Brownian motion in D, i.e., Brownian motion killed at $T(D^c)$.

An excursion law H^x in $D \subset \mathbf{R}^n$ is a σ-finite measure on $(\Omega, \mathcal{F})$ which has the following properties:

(i) $H^x(X_0 \neq x) = 0$,

(ii) H^x is strong Markov for the P_D^x-transition probabilities, i.e.,

$$H^x(a \cdot b(\theta_T)) = H^x\left(a \cdot P_D^{X(T)}(b)\right)$$

for all stopping times $T > 0$, nonnegative and $\mathcal{F}$-measurable b, and nonnegative and $\mathcal{F}_T$-measurable a.

If $D \subset \mathbf{R}^n$ is a Lipschitz domain and $x \in \partial D$ then there exists an excursion law H^x in D.

The following is a version of the exit system theorem. See Maisonneuve (1970) for more details on exit systems and see Revuz (1970) or Williams (1979) for the definition and properties of continuous additive functionals (CAF's).

Suppose that $D \subset \mathbf{R}^n$ is a Lipschitz domain and let μ denote the surface area measure on ∂D. Let L be the CAF of the Brownian motion X (with associated probability measures $\{P^x, x \in \mathbf{R}^n\}$), whose Revuz measure (relative to Lebesgue measure as invariant measure) is given by μ , i.e.,

$$\mu(A) = \lim_{t\downarrow 0} \frac{1}{t} E^v \left[\int_0^t 1_A(X_s) dL_s\right],$$

for all Borel sets $A \subset \mathbf{R}^n$, where v denotes Lebesgue measure on $\mathbf{R}^n$. Fix some nonpolar compact set $B \subset D$. For μ-almost all points $x \in \partial D$, the unit inward normal vector N_x is well defined and $\lim_{\varepsilon\to 0}\varepsilon^{-1}P_D^{x+\varepsilon N_x}(T_B<\infty)$ exists. For such x let H^x be the excursion law in D with the property that $H^x(T_B<\infty)$ is equal to the above limit. For all other x, let $H^x=0$. Then the pair (dL,H) is an exit system in D in the following sense.

For u such that $X_u\in\partial D$ let e_u be the excursion of X in D i.e.,

$$e_u(t)=\begin{cases} X(u+t) & \text{if } \inf\{s>u: X_s\in D^c\}>u+t,\\ \delta & \text{otherwise.}\end{cases}$$

For u such that $X_u\notin\partial D$, define $e_u\equiv\delta$. Then (Burdzy (1987), Theorem 7.2),

$$E^{\cdot}\left(\sum_{0<u<\infty} Z_u\cdot(f\circ e_u)\right)=E^{\cdot}\left(\int_0^\infty Z_s H^{X(s)}(f)dL_s\right) \tag{2.1}$$

for all universally measurable functions f on Ω which vanish on excursions $e_u\equiv\delta$ and nonnegative $\mathcal{F}_t$-predictable processes Z.

3. Some explicit formulas for excursions in a half-space.

Let $D_*=\{x\in\mathbf{R}^n: x_n>0\}$. There exists a unique excursion law H_*^0 in D_* such that $H_*^0(T_B<\infty)=1$ where $B=\{x\in D_*: x_n=1\}$ (see Burdzy (1987) Theorem 3.1). Denote $S_u=\{x\in D_*:|x|=1\}, S_\ell=\{x\in\partial D_*:|x|<1\}, S=\overline{S_u}\cup\overline{S_\ell}$, and $T=\min(T_S,R)$. In the right members below, the symbol dx will denote the differential of Lebesgue measure in $\mathbf{R}^n$, $d\tilde{x}$ will denote the differential of $(n-1)$-dimensional Lebesgue measure (surface measure) on the hyperplane ∂D_*, and $d\sigma=d\sigma(x)$ will denote the differential of surface area measure on the semisphere S_u.

THEOREM 3.1.

(i) $H_*^0(X_t\in dx)=2^{-(n-2)/2}\pi^{-n/2}t^{-(n+2)/2}x_n e^{-|x|^2/(2t)}dx$ *for* $t>0$, $x\in D_*$,

(ii) $H_*^0(|X_t|\in dr)=2^{-(n-2)/2}\pi^{-1/2}(\Gamma((n+1)/2))^{-1}t^{-(n+2)/2}r^n e^{-r^2/(2t)}dr$ for $t>0$, $r>0$,

(iii) $H_*^0(R \in dt, X(R-) \in dx) = (2\pi)^{-n/2} t^{-(n+2)/2} e^{-|x|^2/(2t)} dt d\tilde{x}$ for $t > 0$, $x \in \partial D_*$,

(iv) $H_*^0(R \in dt) = (2\pi t^3)^{-1/2}\, dt$ for $t > 0$,

(v) $H_*^0(R > t) = 2^{1/2}(\pi t)^{-1/2}$ for $t > 0$,

(vi) $H_*^0(X(R-) \in dx) = \Gamma(n/2)\pi^{-n/2}|x|^{-n} d\tilde{x}$ for $x \in \partial D_*$,

(vii) $H_*^0(|X(R-)| > r) = 2\pi^{-1/2}[\Gamma(n/2)/\Gamma((n-1)/2)]r^{-1}$ for $r > 0$,

(viii) $H_*^0\left(\sup_{t \in (0,R)} |X_t| \in dr\right) = 2\pi^{-1/2}[\Gamma((n+2)/2)/\Gamma((n+1)/2)]r^{-2} dr$ for $r > 0$,

(ix) $H_*^0\left(\sup_{t \in (0,R)} |X_t| > r\right) = 2\pi^{-1/2}[\Gamma((n+2)/2)/\Gamma((n+1)/2)]r^{-1}$ for $r > 0$,

(x) $H_*^0(X(T-) \in dx) = \Gamma(n/2)\pi^{-n/2}\left(|x|^{-n} - 1\right) d\tilde{x}$ for $x \in S_\ell$,

(xi) $H_*^0(X(T-) \in dx) = 2\Gamma((n+2)/2)\pi^{-n/2} x_n d\sigma$ for $x \in S_u$,

(xii) $H_*^0(X(T-) \in S_u) = 2\pi^{-1/2}[\Gamma((n+2)/2)/\Gamma((n+1)/2)]$,

(xiii) *The random variables T and $X(T-)$ are conditionally independent under H_*^0 given $\{X(T-) \in S_u\}$. The H_*^0-distribution of T given $\{X(T-) \in S_u\}$ is the same as the distribution of the hitting time of the unit sphere $\{x \in \mathbf{R}^{n+2} : |x| = 1\}$ by the $(n+2)$-dimensional Brownian motion starting at 0.*

Proof: The proofs of parts (i), (iii), (iv), (vi), (viii), (x)-(xiii) were given in Burdzy (1987), Theorem 5.1. Parts (ii), (v), (vii) and (ix) are straightforward consequences of (i), (iv), (vi) and (viii). ∎

We would like to use this occasion to present some formulas for h-processes, related to excursion laws. Let $P_{D_*}^{y,x}$ denote the distribution of the h-process (i.e. conditioned Brownian motion) in D_* which starts at y and converges to x; $E_{D_*}^{y,x}$ will denote the corresponding expectation. See Doob (1984) for the definition of an h-process.

THEOREM 3.2.

(i) $P_{D_*}^{0,x}(R \in dt) = 2^{-n/2}(\Gamma(n/2))^{-1}|x|^n t^{-(n+2)/2} e^{-|x|^2/(2t)}dt$ for $t > 0, x \in \partial D_*$,

(ii) $E_{D_*}^{0,x}(R) = \begin{cases} |x|^2/(n-2) & \text{if } n \geq 3, \\ \infty & \text{if } n = 2, \end{cases}$ for $x \in \partial D_*$,

(iii) $P_{D_*}^{0,x}\left(\sup_{t\in(0,R)} |X_t| \in dr\right) = n|x|^n r^{-(n+1)}dr$ for $r > |x|$, $x \in \partial D_*$,

(iv) $P_{D_*}^{0,x}\left(\sup_{t\in(0,R)} |X_t| > r\right) = (|x|/r)^n$ for $r \geq |x|, x \in \partial D_*$.

Proof: Let $T_\varepsilon = \min(\varepsilon, \inf\{t > 0 : |X_t| = \varepsilon\})$ and apply the strong Markov property at T_ε to obtain for $t > \varepsilon$,

$$(3.1) \qquad P_{D_*}^{0,x}(R \in dt) = \int_0^t \int_{D_*} P_{D_*}^{y,x}(R \in dt - s) P_{D_*}^{0,x}(T_\varepsilon \in ds, X(T_\varepsilon) \in dy).$$

Suppose that the following limit exists

$$(3.2) \qquad \lim_{\substack{y\to 0 \\ y\in D_* \\ s\to 0 \\ s>0}} P_{D_*}^{y,x}(R \in dt - s).$$

Then (3.1) shows that $P_{D_*}^{0,x}(R \in dt)$ is equal to the limit in (3.2).

Observe that

$$(3.3) \qquad P_{D_*}^{y,x}(R \in dt) = \frac{P_{D_*}^{y}(R \in dt, X(R-) \in dx)}{P_{D_*}^{y}(X(R-) \in dx)}.$$

The hitting time of 0 by the 1-dimensional Brownian motion X_n starting from y_n has the density $y_n\left(2\pi t^3\right)^{-1/2} \exp\left(-y_n^2/(2t)\right)$ for $t > 0$. This is the $P_{D_*}^y$-density of R. Given $\{R = t\}$, the $P_{D_*}^y$-distribution of $X(R-)$ is normal with the density

$$(2\pi t)^{-(n-1)/2} \exp(-|\widetilde{x} - \widetilde{y}|^2/(2t))$$

for $x \in \partial D_*$. Multiply the last two formulas to obtain the numerator in (3.3). The denominator is obtained by integration of the numerator over t. It remains to take the limit, as indicated in (3.2) to obtain part (i) of the theorem. Part (ii) follows from (i) by integration.

Although parts (iii) and (iv) may be obtained in a similar, elementary but tedious way, let us point out that using the notation of Theorem 3.1,

$$P_{D_*}^{0,x}\left(\sup_{t\in(0,R)}|X_t|<1\right)=\frac{H_*^0(X(T-)\in dx)}{H_*^0(X(R-)\in dx)}$$

for $x \in \partial D_*, |x| < 1$. Then Theorem 3.1 (vi) and (x) and scaling can be used to obtain (iii) and (iv). ∎

Remark 3.1: The above formulas should be compared with (8.1)-(8.3) of Hsu (1986), although the normalizing constants are not the same.

4. Convergence of excursion laws. Denote $B_1(r) = \{y \in \mathbf{R}^n : |y| \geq r\}$, $B_2(r,v) = \{y \in \mathbf{R}^n : y \cdot v \geq r\}$, where v is a vector in $\mathbf{R}^n$ satisfying $|v| = 1$, and $y \cdot v$ stands for the scalar product. For a set $D \subset \mathbf{R}^n$ let $B_3(r,D) = \{y \in \mathbf{R}^n : \text{dist}\,(y, \partial D) \geq r\}$. Consider the following events:

$$A_1(t,r) = \{|X_t| > r\},$$

$$A_2(t,r) = \{R > t, |X(R-)| > r\},$$

$$A_3(t) = \{R > t\},$$

$$A_4(r) = \{|X(R-)| > r\},$$

$$A_5(r) = \{T(B_1(r)) < \infty\},$$

$$A_6(r,v) = \{T(B_2(r,v)) < \infty\},$$

$$A_7(r,D) = \{T(B_3(r,D)) < \infty\}.$$

Suppose that $D \subset \mathbf{R}^n$ is a Greenian domain. Let $f_k(x,t) = P_D^x(A_k)$ for $1 \leq k \leq 7$. Of course, f_k depends also on D and r.

For $\varepsilon > 0$, let $B(x, \varepsilon) = \{y \in \mathbf{R}^n : |x - y| < \varepsilon\}$ and $T_{x,\varepsilon} = \min\left(\varepsilon^2, T(\partial B(x, \varepsilon))\right)$. Apply the strong Markov property at $T_{x,\varepsilon}$ to see that

$$f_k(x, t) = \int_0^t \int_D f_k(y, t - u) P_D^x \left(X\left(T_{x,\varepsilon}\right) \in dy, T_{x,\varepsilon} \in du\right)$$

for $1 \le k \le 7$, provided $\varepsilon^2 < t, B(x, \varepsilon) \subset D$ and $B(x, \varepsilon) \subset D \setminus B_{k-4}$ for $k = 5, 6, 7$. This averaging property means that the functions f_k are parabolic in $D \times (0, \infty)$ for $k = 1, 2, 3, 4$ and in $(D \setminus B_{k-4}) \times (0, \infty)$ for $k = 5, 6, 7$ (see Doob (1984), p.276).

The next proposition contains a comparison result for certain distributions of Brownian motion in a Lipschitz domain that is locally approximable by a half-space.

PROPOSITION 4.1. *Let $D_* = \{y \in \mathbf{R}^n : y \cdot v > 0\}$ for some $v \in \mathbf{R}^n$ satisfying $|v| = 1$, and let f_k^* and f_k correspond to domains D_* and D (the last one is described below). For positive λ, r, u, α and ε, there exist $\rho = \rho(n, \lambda, r, \varepsilon, u, \alpha) < \min(\sqrt{u}, \alpha)$ and $\varepsilon_1 = \varepsilon_1(\rho)$ with the following property.*

Suppose that φ is a Lipschitz function with constant λ and D is a domain such that

$$\{y \in D : |y| < 1/\varepsilon_1\} = \{y \in D : |y| < 1/\varepsilon_1, y_n > \varphi(\widetilde{y})\}.$$

Assume that

$$\{y \in \partial D : |y| < 1/\varepsilon_1\} \subset \{y \in \mathbf{R}^n : |y| < 1/\varepsilon_1, y \cdot v \in (-\varepsilon_1, \varepsilon_1)\}.$$

Then

$$\frac{f_k\left(\underline{A}_{\rho/2}(0, u)\right)}{f_k\left(\overline{A}_{\rho/2}(0, u)\right)} \ge 1/2, \tag{4.1}$$

$$\text{(4.2)} \qquad \frac{f_k^*\left(\underline{A}_{\rho/2}(0,u)\right)}{f_k^*\left(\overline{A}_{\rho/2}(0,u)\right)} \geq 1/2$$

and

$$\text{(4.3)} \qquad \frac{f_k((\widetilde{0},\rho/32),u)}{f_k^*((\widetilde{0},\rho/32),u)} \in (1-\varepsilon, 1+\varepsilon)$$

for $1 \leq k \leq 7$.

Remark 4.1: The fraction $\rho/32$ appears here because it is used in later estimates.

Proof: First consider (4.2). It obviously holds for $k = 4, 5, 6$ and 7 since in these cases f_k^* does not depend on t.

Recall the following explicit formulas from Section 3 and the proof of Theorem 5.1 of Burdzy (1987). Here we use a coordinate system which makes D_* the half-space $\{y \in \mathbf{R}^n : y_n > 0\}$.

$$P_{D_*}^x(X_t \in dy) = (2\pi t)^{-n/2} \exp\left(-|x-y|^2/\,2t\right)\left(1 - \exp\left(-2x_n y_n/\,t\right)\right) dy,$$

$$P_{D_*}^x(R \in dt) = \left(2\pi t^3\right)^{-1/2} x_n \exp\left(-x_n^2/\,2t\right) dt,$$

$$P_{D_*}^x(R \in dt, X(R-) \in dy)$$
$$= \left(2\pi t^3\right)^{-1/2} x_n \exp\left(-x_n^2/\,2t\right)(2\pi t)^{-(n-1)/2} \exp(-|\widetilde{x} - \widetilde{y}|^2/\,2t) dt\, d\widetilde{y}.$$

Given these explicit formulas, it is elementary to check that for fixed r and u,

$$\frac{f_k^*\left(\underline{A}_q(0,u)\right)}{f_k^*\left(\overline{A}_q(0,u)\right)} \to 1$$

as $q \to 0$, for $k = 1, 2, 3$. Choose $\rho \in (0, \min(\sqrt{u}, \alpha))$ so that (4.2) holds even with 1/2 replaced by 3/4.

Now for the proof of (4.1) and (4.3), let

$$D^m = \{y \in \mathbf{R}^n : |y| < 1/\varepsilon_1, y \cdot v > \varepsilon_1\}$$

and

$$D^M = \{y \in \mathbf{R}^n : |y| > 1/\varepsilon_1 \text{ or } y \cdot v > -\varepsilon_1\}.$$

Observe that $D^m \subset D_* \subset D^M$ and $D^m \subset D \subset D^M$. Let f_k^m and f_k^M correspond to D^m and D^M. The continuity of probability implies that for a fixed $x \in D_*$ and $t > 0$,

$$f_k^m(x,t) - f_k^M(x,t) \to 0 \tag{4.4}$$

as $\varepsilon_1 \to 0$ for $1 \leq k \leq 7$. Since f_k is a monotone function of D for $k = 1,3,5,6$, the formula (4.4) implies that for these values of k and fixed $x \in D_*$ and $t > 0$,

$$f_k^*(x,t) - f_k(x,t) \to 0 \text{ as } \varepsilon_1 \to 0. \tag{4.5}$$

It is easy to see that for any fixed x, the P^x probability of the union of the events

$$\{[T_{\partial D} > t \text{ and } |X(T_{\partial D})| > r] \text{ and } [T_{\partial D_*} \leq t \text{ and } |X(T_{\partial D_*})| \leq r]\}$$

and

$$\{[T_{\partial D_*} > t \text{ and } |X(T_{\partial D_*})| > r] \text{ and } [T_{\partial D} \leq t \text{ or } |X(T_{\partial D})| \leq r]\}$$

tends to zero as $\varepsilon_1 \to 0$. It follows that

$$f_k^*(x,t) - f_k(x,t) \to 0 \text{ as } \varepsilon_1 \to 0 \tag{4.6}$$

for $k = 2$ and, for similar reasons, for $k = 4$ and 7.

Let $(x,t) = ((\tilde{0},\rho/32),u)$ in (4.5) and (4.6) to see that (4.3) holds for suitably small ε_1.

When $\underline{A}_{\rho/2}(0,u)$ and $\overline{A}_{\rho/2}(0,u)$ are substituted for (x,t) in (4.5) and (4.6) then these formulas, together with (4.2) (recall this holds with 3/4 in place of 1/2), imply that (4.1) holds for small ε_1. ∎

Let D be a Lipschitz domain in $\mathbf{R}^n$. Fix some $z^0 \in D$ and let

$$B = \left\{ y \in D : G_D\left(y, z^0\right) \geq 1 \right\}.$$

THEOREM 4.1. *For each $\varepsilon > 0$ there exists $\varepsilon_1 = \varepsilon_1(\varepsilon, n, \lambda)$ such that the following holds.*

Suppose that $a > 0$ satisfies

$$\{y \in D : |y| < a/\varepsilon_1\} = \{y \in D : |y| < a/\varepsilon_1, y_n > \varphi(\widetilde{y})\} \subset D \setminus B$$

where φ is a Lipschitz function with constant λ, $\varphi(0) = 0$, and for some v with $|v| = 1$,

$$\{y \in \partial D : |y| < a/\varepsilon_1\} \subset \{y \in \mathbf{R}^n : |y| < a/\varepsilon_1, y \cdot v \in (-a\varepsilon_1, a\varepsilon_1)\}.$$

Let H^0 be an excursion law in D with $H^0(T_B < \infty) \in (0, \infty)$. In the definitions of events $A_k, 1 \leq k \leq 7$, let $r = a, t = a^2$, and v and D be as above. Then

$$H^0(A_k) \cdot \frac{G_D\left(av, z^0\right)}{H^0(T_B < \infty)} \in (d_k(1-\varepsilon), d_k(1+\varepsilon))$$

for $1 \leq k \leq 7$. Here the d_k's are given by

$$d_1 = \int_1^\infty 2^{-(n-2)/2}\pi^{-1/2}(\Gamma((n+1)/2))^{-1}s^n e^{-s^2/2}ds,$$

$$d_2 = \int_1^\infty \int_1^\infty 2^{-(n-2)/2}\pi^{-1/2}(\Gamma((n-1)/2))^{-1}s^{-(n+2)/2}r^{n-2}e^{-r^2/(2s)}dsdr,$$

$$d_3 = (2/\pi)^{1/2},$$

$$d_4 = 2\pi^{-1/2}[\Gamma(n/2)/\Gamma((n-1)/2)],$$

$$d_5 = 2\pi^{-1/2}[\Gamma((n+2)/2)/\Gamma((n+1)/2)],$$

$$d_6 = d_7 = 1.$$

Proof: Denote $f_8^*(x,t) = x \cdot v$ for $x \in D_* \stackrel{\mathrm{df}}{=} \{y \in \mathbb{R}^n : y \cdot v > 0\}, t > 0$, and $f_8(x,t) = aG_D(x, z^0) / G_D(av, z^0)$ for $x \in D, t > 0$. Let $\varepsilon_2, \varepsilon_3 > 0$ be small constants which will be specified later.

By the proof of Proposition 4.1 and scaling it is possible to choose $\rho < \min(1, 32\varepsilon_2/a)$ and $\varepsilon_1 > 0$ small enough so that for $1 \le k \le 7$,

$$\frac{f_k\left(\underline{A}_{a\rho/2}(0,a^2)\right)}{f_k\left(\overline{A}_{a\rho/2}(0,a^2)\right)} \ge 1/2, \tag{4.7}$$

$$\frac{f_k^*\left(\underline{A}_{a\rho/2}(0,a^2)\right)}{f_k^*\left(\overline{A}_{a\rho/2}(0,a^2)\right)} \ge 1/2 \tag{4.8}$$

and

$$\frac{f_k((\widetilde{0}, a\rho/32), a^2)}{f_k^*((\widetilde{0}, a\rho/32), a^2)} \in (1-\varepsilon_2, 1+\varepsilon_2). \tag{4.9}$$

We have obviously

$$\frac{f_8^*(\underline{A}_{a\rho/2}(0,a^2))}{f_8^*(\overline{A}_{a\rho/2}(0,a^2))} = 1 \tag{4.10}$$

and

$$\frac{f_8(\underline{A}_{a\rho/2}(0,a^2))}{f_8(\overline{A}_{a\rho/2}(0,a^2))} = 1. \tag{4.11}$$

The Green function in a half-space behaves near the boundary like a linear function. It is easy to see that for fixed x and z^0, $G_D(x,z^0) \to G_{D_*}(x,z^0)$ as $\varepsilon_1 \to 0$ and, therefore, for ε_1 small enough

$$\frac{f_8^*((\widetilde{0}, a\rho/32), a^2)}{f_8((\widetilde{0}, a\rho/32), a^2)} = \frac{((\widetilde{0}, a\rho/32)\cdot v) G_D(av, z^0)}{G_D((\widetilde{0}, a\rho/32), z^0) a} \in (1-\varepsilon_2, 1+\varepsilon_2)$$

and this together with (4.9) implies that

$$\frac{f_k((\widetilde{0},a\rho/32),a^2)\, f_8^*((\widetilde{0},a\rho/32),a^2)}{f_k^*((\widetilde{0},a\rho/32),a^2)\, f_8((\widetilde{0},a\rho/32),a^2)} \in ((1-\varepsilon_2)^2,(1+\varepsilon_2)^2). \tag{4.12}$$

Now Corollary 6.1 will be applied. Its assumptions are satisfied due to (4.7)-(4.12). Let ε_2 (and consequently ε_1) be so small that (6.21) holds with $c = \varepsilon_3$ for the functions f_k, f_k^*, f_8 and f_8^* i.e.,

$$\lim_{\substack{(x,t)\to(0,a^2)\\ x\in D\\ t>0}} \frac{f_k(x,t)}{f_8(x,t)} \cdot \lim_{\substack{(y,u)\to(0,a^2)\\ y\in D_*\\ u>0}} \frac{f_8^*(y,u)}{f_k^*(y,u)} \in (1-\varepsilon_3, 1+\varepsilon_3).$$

Define d_k by declaring that the second limit in the above formula is equal to a/d_k.

Choose $0 < \varepsilon_4 < a$ so small that $\{x : |x| \le \varepsilon_4\} \cap B = \emptyset$ and

$$\frac{f_k(x,t)a}{f_8(x,t)d_k} \in (1-2\varepsilon_3, 1+2\varepsilon_3) \tag{4.13}$$

for $x \in D$, $|x| < \varepsilon_4$, $|t-a^2| < \varepsilon_4^2$. Denote

$$T = \min\left(\varepsilon_4^2, T(\{y \in \mathbb{R}^n : |y| \ge \varepsilon_4\})\right).$$

Apply the strong Markov property at T, and use (4.13) together with the definition of f_8, to see that

$$\begin{aligned}
H^0(A_k) &= \int_0^{\varepsilon_4^2}\int_D f_k\left(y,a^2-s\right) H^0(X(T)\in dy,\ T\in ds)\\
&\le \int_0^{\varepsilon_4^2}\int_D (d_k/a)(1+2\varepsilon_3)\, f_8\left(y,a^2-s\right) H^0(X(T)\in dy,\ T\in ds)\\
&= (d_k/a)(1+2\varepsilon_3)\, a/G_D\left(av,z^0\right) \int_0^{\varepsilon_4^2}\int_D G_D\left(y,z^0\right) H^0(X(T)\in dy,\ T\in ds)\\
&= d_k\left(1+2\varepsilon_3\right)/G_D\left(av,z^0\right) \int_0^{\varepsilon_4^2}\int_D P^y\left(T_B<\infty\right) H^0(X(T)\in dy,\ T\in ds)\\
&= \left[d_k\left(1+2\varepsilon_3\right)/G_D\left(av,z^0\right)\right] H^0\left(T_B<\infty\right).
\end{aligned}$$

To obtain the second last equality we have used the fact that $G_D(y, z^0)$ and $P^y(T_B < \infty)$ are equal on $D \setminus B$, both being harmonic there with the same boundary values and vanishing at infinity if D is unbounded. Analogously,

$$H^0(A_k) \geq \left[d_k(1 - 2\varepsilon_3)/G_D(av, z^0)\right] H^0(T_B < \infty).$$

Set ε_3 to $\varepsilon/2$ to obtain the desired result. As for the d_k's, note that $d_k/a = H^0_*(A_k)$ and apply Theorem 3.1 to find their values. ∎

5. Local time representations. Let D be a Lipschitz domain in $\mathbf{R}^n$. Recall the definition of an excursion e_t of X in D and of the local time L of the Brownian motion X, $\{P^x, x \in \mathbf{R}^n\}$, from Section 2.

THEOREM 5.1. *Suppose $\varepsilon_1 > 0$ and $h_k : \mathbf{R}^{n-1} \to \mathbf{R}$, $k = 1, 2$, are Lipschitz functions with constant $\lambda > 0$, satisfying $h_1 \leq 0 \leq h_2$ and*

$$\int_{\{x \in \mathbf{R}^{n-1}: |x| < 1\}} |h_k(x)|\, |x|^{-n} dx < \infty, \quad k = 1, 2. \tag{5.1}$$

Further suppose that for each $x \in \partial D$ there is a Lipschitz function $\varphi_x : \mathbf{R}^{n-1} \to \mathbf{R}$ with constant λ such that $\varphi_x(0) = 0$, and in a suitable orthonormal coordinate system $CS(x)$ where $x = 0$, we have

$$\{y \in \partial D : |y| < \varepsilon_1\} = \{y \in \mathbf{R}^n : |y| < \varepsilon_1, y_n = \varphi_x(\widetilde{y})\}$$
$$\subset \{y \in \mathbf{R}^n : |y| < \varepsilon_1, h_1(\widetilde{y}) \leq y_n \leq h_2(\widetilde{y})\}.$$

Let $N_t^k(\varepsilon)$ be the number of excursions e_s of X in D such that $s \leq t$ and $e_s \in A_k$. Here the A_k are the events defined in Section 4; in their definition we take $r = \varepsilon, t = \varepsilon^2$, and v to be the inward unit normal vector at $e_s(0)$, if it exists, otherwise $v = (1, 0, \ldots, 0)$.

Then for each $t > 0, 1 \leq k \leq 7$,

$$\lim_{\varepsilon \to 0} \varepsilon \cdot N_t^k(\varepsilon)/d_k = L_t$$

where the convergence holds in P^x-probability (for each $x \in \mathbf{R}^n$). See Theorem 4.1 for formulas for the d_k.

Proof: First we prove an asymptotic comparison result (5.14) for the Green function in D. Fix some $x \in \partial D$ and use the coordinate system $CS(x)$. Let

$$\begin{aligned} D_k &= \{y \in \mathbf{R}^n : |y| < \varepsilon_1, y_n > h_k(\widetilde{y})\}, k = 1,2, \\ z^0 &= (\widetilde{0}, \varepsilon_1/2), \\ V &= \{y \in D : y = (\widetilde{0}, b), b > 0\}. \end{aligned}$$

Theorem 4.2 of Burdzy and Williams (1986) implies in view of (5.1) that

$$\lim_{\substack{z \to 0 \\ z \in V}} G_{D_k}(z^0, z)/|z| = q_k \in (0, \infty), \qquad k = 1,2. \tag{5.2}$$

Let $\varepsilon_2 > 0$; its value will be specified later. Find $\varepsilon_3 > 0$ so small that one has (use (5.2))

$$G_{D_1}(z^0, z)/(|z| q_1) \in (1 - \varepsilon_2, 1 + \varepsilon_2) \tag{5.3}$$

and

$$\frac{G_{D_1}(z^0, z)}{G_{D_2}(z^0, z)} \in \left(\frac{q_1}{q_2}(1 - \varepsilon_2), \frac{q_1}{q_2}(1 + \varepsilon_2) \right) \tag{5.4}$$

for $z \in V, |z| < \varepsilon_3 \varepsilon_1$.

It follows from the elliptic boundary Harnack principle (see the version presented in Theorem 2.2 of Burdzy (1987)) that there exists $\varepsilon_4 > 0$ such that if g_1 and g_2 are positive harmonic functions in

$$D_k^a \overset{\text{df}}{=} \{y \in \mathbf{R}^n : |y| < \varepsilon_1 a, y_n > h_k(\widetilde{y})\}, \qquad a < 1/2,$$

which vanish on $\partial D_k^a \setminus D_k$, and $y^m \in D_k^a, |y^m| < \varepsilon_1 a \varepsilon_4, m = 1,2$, then

$$\frac{g_1(y^1)}{g_1(y^2)} \frac{g_2(y^2)}{g_2(y^1)} \in (1 - \varepsilon_2, 1 + \varepsilon_2). \tag{5.5}$$

The constant ε_4 depends only on n, λ and ε_2.

It is elementary to prove that (5.1) implies that

$$\lim_{y\to 0} h_k(y)/|y| = 0, \qquad k = 1, 2. \tag{5.6}$$

Denote $M_k^a = \partial D_k^a \cap D_k$. We can choose a sufficiently small that

$$\frac{P_{D_1^a}^{z^1}\left(T\left(M_1^a-\right)<\infty\right)}{P_{D_2^a}^{z^1}\left(T\left(M_2^a-\right)<\infty\right)} \in (1-\varepsilon_2, 1+\varepsilon_2) \tag{5.7}$$

where $z^1 \in V, \left|z^1\right| < \varepsilon_1 \min(\varepsilon_3, \varepsilon_4 a)/2$. Apply (5.5) to see that

$$\frac{P_{D_1^a}^{z^1}\left(T\left(M_1^a-\right)<\infty\right)}{P_{D_1^a}^{z}\left(T\left(M_1^a-\right)<\infty\right)} \frac{G_{D_1}\left(z^0, z\right)}{G_{D_1}\left(z^0, z^1\right)} \in (1-\varepsilon_2, 1+\varepsilon_2) \tag{5.8}$$

for $z \in D_1, |z| < \varepsilon_1\varepsilon_4 a$ and

$$\frac{P_{D_2^a}^{z}\left(T\left(M_2^a-\right)<\infty\right)}{P_{D_2^a}^{z^1}\left(T\left(M_2^a-\right)<\infty\right)} \frac{G_{D_2}\left(z^0, z^1\right)}{G_{D_2}\left(z^0, z\right)} \in (1-\varepsilon_2, 1+\varepsilon_2)$$

for $z \in D_2, |z| < \varepsilon_1\varepsilon_4 a$. The last two formulas imply that

(5.9)
$$\frac{P_{D_2^a}^{z}\left(T\left(M_2^a-\right)<\infty\right)}{P_{D_1^a}^{z}\left(T\left(M_1^a-\right)<\infty\right)} \frac{P_{D_1^a}^{z^1}\left(T\left(M_1^a-\right)<\infty\right)}{P_{D_2^a}^{z^1}\left(T\left(M_2^a-\right)<\infty\right)} \times$$
$$\times \frac{G_{D_1}\left(z^0, z\right)}{G_{D_2}\left(z^0, z\right)} \frac{G_{D_2}\left(z^0, z^1\right)}{G_{D_1}\left(z^0, z^1\right)} \in \left((1-\varepsilon_2)^2, (1+\varepsilon_2)^2\right)$$

for $z \in V, |z| < \varepsilon_1\varepsilon_4 a$. Let $\varepsilon_5 = \min(\varepsilon_3, \varepsilon_4 a)$. Then (5.4), (5.7) and (5.9) imply

$$\frac{P_{D_2^a}^{z}\left(T\left(M_2^a-\right)<\infty\right)}{P_{D_1^a}^{z}\left(T\left(M_1^a-\right)<\infty\right)} \in \left(\frac{(1-\varepsilon_2)^3}{(1+\varepsilon_2)^2}, \frac{(1+\varepsilon_2)^3}{(1-\varepsilon_2)^2}\right) \tag{5.10}$$

for $z \in V, |z| < \varepsilon_1\varepsilon_5$.

Let $D^a = \{y \in D : |y| < \varepsilon_1 a\}, M^a = \partial D^a \cap D$. Since $D_1^a \subset D^a \subset D_2^a$, one has

$$P_{D_2^a}^{z}\left(T\left(M_2^a-\right)<\infty\right) \le P_{D^a}^{z}\left(T\left(M^a-\right)<\infty\right) \le P_{D_1^a}^{z}\left(T\left(M_1^a-\right)<\infty\right).$$

It follows from this and (5.10) that

$$P^z_{D^a}(T(M^a-)<\infty) \Big/ P^z_{D^a_1}(T(M^a_1-)<\infty) \in \left(\frac{(1-\varepsilon_2)^3}{(1+\varepsilon_2)^2}, \frac{(1+\varepsilon_2)^3}{(1-\varepsilon_2)^2}\right) \tag{5.11}$$

for $z \in V, |z| < \varepsilon_1\varepsilon_5$. Combine (5.3), (5.8) and (5.11) to see that

$$\frac{P^z_{D^a}(T(M^a-)<\infty)}{|z|} \frac{G_{D_1}(z^0,z^1)}{q_1 P^{z^1}_{D^a_1}(T(M^a_1-)<\infty)} \in \left(\frac{(1-\varepsilon_2)^4}{(1+\varepsilon_2)^3}, \frac{(1+\varepsilon_2)^4}{(1-\varepsilon_2)^3}\right) \tag{5.12}$$

for $z \in V, |z| < \varepsilon_1\varepsilon_5$. By (5.5), for any $z^2 \notin D^a$,

$$\frac{G_D(z^2,z)}{G_D(z^2,z^1)} \frac{P^{z^1}_{D^a}(T(M^a-)<\infty)}{P^z_{D^a}(T(M^a-)<\infty)} \in (1-\varepsilon_2, 1+\varepsilon_2),$$

which combined with (5.12) yields

$$\frac{G_D(z^2,z)}{|z|} \frac{|z^1|}{G_D(z^2,z^1)} \in \left(\frac{(1-\varepsilon_2)^9}{(1+\varepsilon_2)^8}, \frac{(1+\varepsilon_2)^9}{(1-\varepsilon_2)^8}\right) \tag{5.13}$$

for $z \in V, \ |z| < \varepsilon_1\varepsilon_5$.

Now, given any $\varepsilon_6 > 0$, by choosing ε_2 sufficiently small, it follows from the above that there exists $\varepsilon_7 > 0$ which depends on n, λ and ε_6 (but not x) such that for any $z^2 \in D, \ |z^2 - x| > \varepsilon_1/2, z^3 \in V, |z^3| \le \varepsilon_7$,

$$\frac{|z^3|}{G_D(z^3,z^2)} \lim_{\substack{z\to 0 \\ z\in V}} \frac{G_D(z^2,z)}{|z|} \in (1-\varepsilon_6, 1+\varepsilon_6). \tag{5.14}$$

The limit above exists according to Theorem 4.2 of Burdzy and Williams (1986) whose assumptions are satisfied due to (5.1).

We now use (5.14) to prove the local time representation result. Recall the definition of an exit system (dL, H) in D, from Section 2. The continuous additive functional L has the surface area measure on ∂D as its Revuz measure.

Fix some $z^2 \in D$ with dist $\left(z^2, \partial D\right) > \varepsilon_1/2$ so that the assumptions of (5.14) are satisfied for every $x \in \partial D$. Let

$$B = \left\{y \in D : G_D\left(z^2, y\right) \geq 1\right\}.$$

Then $G_D\left(z^2, y\right) = P_D^y\left(T_B < \infty\right)$ for $y \in D \backslash B$. It follows from the comments following (5.14) that for each $x \in \partial D$

$$\lim_{\substack{z \to 0 \\ z \in V}} P_D^z\left(T_B < \infty\right)/|z|$$

exists (the formula is expressed in $CS(x)$). The excursion laws are normalized so that

$$H^x\left(T_B < \infty\right) = \lim_{\substack{z \to 0 \\ z \in V}} P_D^z\left(T_B < \infty\right)/|z| \text{ in } CS(x).$$

In view of (5.6), Theorem 4.1 and (5.14) imply that for each $\varepsilon_8 > 0$ one may choose $\varepsilon_9 > 0$ so that for $1 \leq k \leq 7$

$$\text{(5.15)} \qquad H^x\left(A_k\right) \in \left(d_k\left(1 - \varepsilon_8\right)/\varepsilon, d_k\left(1 + \varepsilon_8\right)/\varepsilon\right)$$

if $r = \varepsilon \leq \varepsilon_9$ and $t = \varepsilon^2$ in the definition of A_k.

Denote $\sigma(s) = \inf\{t > 0 : L_t > s\}$. Theorem T4 from Chapter II of Brémaud (1981) and the exit system formula (2.1) imply that for $\varepsilon < \varepsilon_9$ the process $s \to N_{\sigma(s)}^k(\varepsilon)$ is Poisson with a random intensity which by (5.15) is bounded below by $d_k\left(1 - \varepsilon_8\right)/\varepsilon$ and above by $d_k\left(1 + \varepsilon_8\right)/\varepsilon$. When $\varepsilon \to 0$, one may let ε_8 go to 0 as well and for a fixed s, $\varepsilon \cdot N_{\sigma(s)}^k(\varepsilon)/d_k$ converges in probability to s; this may be easily deduced for example from formula (1.9) of Chapter II of Brémaud (1981). It is now elementary to see that $\varepsilon \cdot N_t^k(\varepsilon)/d_k$ converges in probability to L_t, for a fixed t. ∎

Remark 5.1: The above representation theorem works, for example, for $C^{1,\alpha}$ domains with $\alpha > 0$, i.e., for domains which have boundaries represented locally by functions whose first partial derivatives are α-Hölder continuous.

6. A parabolic boundary Harnack principle. The following result is a stronger version of Lemma 2.1 of Burdzy (1987).

In the sequel, inequalities involving zero divisors are to be interpreted as those obtained by multiplication by the divisors.

LEMMA 6.1. *Suppose that $b, c, d \in (0,1)$, and f_1, f_2, g_x, g_y are real-valued, nonnegative measurable functions defined on a set $W = U \cup V$ where U and V are disjoint measurable sets. Let ν be an arbitrary positive measure on W. Assume that*

$$\frac{f_k(v)}{f_{3-k}(v)} \geq c \frac{f_k(w)}{f_{3-k}(w)} \quad \text{for all } v, w \in W,\ k = 1, 2, \tag{6.1}$$

and

$$\frac{g_x(v)}{g_y(v)} \geq d\, \frac{g_x(w)}{g_y(w)} \quad \text{for all } v, w \in V. \tag{6.2}$$

Let

$$h_k(z) = \int_W f_k g_z d\nu \overset{\text{df}}{=} \int_W f_k(v) g_z(v) d\nu(v)$$

and

$$\widetilde{h}_k(z) = \int_V f_k g_z d\nu \overset{\text{df}}{=} \int_V f_k(v) g_z(v) d\nu(v)$$

for $k = 1, 2$ and $z = x, y$. Suppose that

$$\infty > \widetilde{h}_k(z) \geq b h_k(z) \tag{6.3}$$

for $k = 1, 2$ and $z = x, y$. Then

$$\frac{h_2(x)}{h_1(x)} \geq \frac{h_2(y)}{h_1(y)} \left[c + b^2 d^2 (1 - c) \right].$$

Proof: Choose $v_0 \in V$ so that

$$g_x(v_0) \,/\, g_y(v_0) \overset{\text{df}}{=} q \in (0, \infty).$$

It is easy to see that if such a v_0 does not exist then the lemma trivially holds. By (6.2), $g_x(v) \geq dq\, g_y(v)$ for all $v \in V$. It follows that $\tilde{g}(v) \geq 0$ for all $v \in V$, where

$$\tilde{g}(v) \stackrel{\text{df}}{=} g_x(v) - dq\, g_y(v).$$

By (6.2), $g_y(v) \geq g_x(v)d/q$. Apply this inequality to see that

$$\tilde{h}_1(y) = \int_V f_1 g_y d\nu \geq \int_V f_1 g_x (d/q) d\nu = \tilde{h}_1(x) d/q. \tag{6.4}$$

By (6.1),

$$f_2(v) f_1(w) \geq c f_1(v) f_2(w) \quad \text{for all } v, w \in W. \tag{6.5}$$

Hence,

$$\left(\int_V f_2 \tilde{g} d\nu\right)\left(\int_V f_1 g_y d\nu\right) \geq c\left(\int_V f_1 \tilde{g} d\nu\right)\left(\int_V f_2 g_y d\nu\right),$$

or equivalently

$$\tilde{h}_1(y) \int_V f_2 \tilde{g} d\nu \geq c\tilde{h}_2(y) \int_V f_1 \tilde{g} d\nu. \tag{6.6}$$

In an analogous way, we obtain from (6.5) the following inequalities.

$$\tilde{h}_1(y) \int_U f_2 g_x d\nu \geq c\tilde{h}_2(y) \int_U f_1 g_x d\nu \tag{6.7}$$

and

$$h_2(x) \int_U f_1 g_y d\nu \geq c h_1(x) \int_U f_2 g_y d\nu. \tag{6.8}$$

By the definition of $\widetilde{g}$, (6.6) and (6.4),

$$\begin{aligned}
\widetilde{h}_2(x)\widetilde{h}_1(y) &= \left(\int_V f_2\widetilde{g}d\nu + dq\int_V f_2 g_y d\nu\right)\widetilde{h}_1(y) \\
&= \left(\int_V f_2\widetilde{g}d\nu + dq\widetilde{h}_2(y)\right)\widetilde{h}_1(y) \\
&\geq \left(c\int_V f_1\widetilde{g}d\nu + dq\widetilde{h}_1(y)\right)\widetilde{h}_2(y) \\
&= \left(c\left(\int_V f_1 g_x d\nu - dq\int_V f_1 g_y d\nu\right) + dq\widetilde{h}_1(y)\right)\widetilde{h}_2(y) \qquad (6.9)\\
&= (c\widetilde{h}_1(x) + dq(1-c)\widetilde{h}_1(y))\widetilde{h}_2(y) \\
&\geq (c\widetilde{h}_1(x) + d^2(1-c)\widetilde{h}_1(x))\widetilde{h}_2(y) \\
&= (c + d^2(1-c))\widetilde{h}_1(x)\widetilde{h}_2(y).
\end{aligned}$$

By (6.7), (6.9) and (6.3),

$$\begin{aligned}
h_2(x)\widetilde{h}_1(y) &= \left(\int_U f_2 g_x d\nu + \int_V f_2 g_x d\nu\right)\widetilde{h}_1(y) \\
&= \left(\int_U f_2 g_x d\nu + \widetilde{h}_2(x)\right)\widetilde{h}_1(y) \\
&\geq \left(c\int_U f_1 g_x d\nu + (c + d^2(1-c))\widetilde{h}_1(x)\right)\widetilde{h}_2(y) \\
&= (ch_1(x) + d^2(1-c)\widetilde{h}_1(x))\widetilde{h}_2(y) \\
&\geq (ch_1(x) + d^2(1-c)bh_1(x))\widetilde{h}_2(y) \\
&= (c + bd^2(1-c))h_1(x)\widetilde{h}_2(y).
\end{aligned}$$

Then, by the last inequality, (6.8) and (6.3), we obtain

$$\begin{aligned}
h_2(x)h_1(y) &= \left(\int_U f_1 g_y d\nu + \int_V f_1 g_y d\nu\right)h_2(x) \\
&= \left(\int_U f_1 g_y d\nu + \widetilde{h}_1(y)\right)h_2(x) \\
&\geq \left(c\int_U f_2 g_y d\nu + (c + bd^2(1-c))\widetilde{h}_2(y)\right)h_1(x) \\
&= (ch_2(y) + bd^2(1-c)\widetilde{h}_2(y))h_1(x) \\
&\geq (ch_2(y) + b^2d^2(1-c)h_2(y))h_1(x) \\
&= (c + b^2d^2(1-c))h_2(y)h_1(x). \quad \blacksquare
\end{aligned}$$

Suppose that D is a Lipschitz domain and moreover $D = \{x \in \mathbb{R}^n : x_n > \varphi(\widetilde{x})\}$ where $|\varphi(\widetilde{x}) - \varphi(\widetilde{y})| \leq \lambda|\widetilde{x} - \widetilde{y}|$ and $\varphi(\widetilde{0}) = 0$.

Recall the definitions of $\Psi, \underline{A}, \overline{A}$ and Δ from Section 2.

THEOREM 6.1. *There exists a function $c = c(a,b,n,\lambda,\varepsilon)$, $a, b, \lambda, \varepsilon > 0, n \geq 2$, with the following properties.*

(i) $c \in (0,1)$, c is decreasing in ε and increasing in a and b.

(ii) For fixed a, b, n and λ,

$$\lim_{\varepsilon \downarrow 0} c(a,b,n,\lambda,\varepsilon) = 1.$$

(iii) Let $s > 0$ and $0 < r < \sqrt{s}$ and suppose that $f_1(x,t)$ and $f_2(x,t)$ are positive and parabolic in $\Psi_r(0,s)$ and they vanish continuously on $\Delta_r(0,s)$. Then, for $(x,t),(y,u) \in \Psi_\varepsilon(0,s)$, $\varepsilon < r/16, k = 1,2$, we have

$$\frac{f_k(x,t)}{f_{3-k}(x,t)} \geq \frac{f_k(y,u)}{f_{3-k}(y,u)} \cdot c$$

where

$$c = c\left(\frac{f_1\left(\underline{A}_{r/2}(0,s)\right)}{f_1\left(\overline{A}_{r/2}(0,s)\right)}, \frac{f_2\left(\underline{A}_{r/2}(0,s)\right)}{f_2\left(\overline{A}_{r/2}(0,s)\right)}, n, \lambda, \varepsilon\right).$$

Proof: We will suppress $(0,s)$ in the notation i.e., $\Psi_\rho = \Psi_\rho(0,s), \underline{A}_\rho = \underline{A}_\rho(0,s)$ etc. We first establish some inequalities so that we can apply Lemma 6.1 and then we use induction to obtain the theorem.

By Theorem 1.6 (see also inequality (1.28)) of Fabes et al. (1986) we have for $k = 1,2$, and (x,t), $(y,u) \in \Psi_{r/16}$,

$$\frac{f_k(x,t)}{f_{3-k}(x,t)} \geq \frac{f_k(y,u)}{f_{3-k}(y,u)} \frac{f_1\left(\underline{A}_{r/2}\right)}{f_1\left(\overline{A}_{r/2}\right)} \frac{f_2\left(\underline{A}_{r/2}\right)}{f_2\left(\overline{A}_{r/2}\right)} c_1^2 \tag{6.10}$$

where $c_1 = c_1(\lambda, n) > 0$. Now let

$$c_2 = \frac{f_1\left(\underline{A}_{r/2}\right)}{f_1\left(\overline{A}_{r/2}\right)} \frac{f_2\left(\underline{A}_{r/2}\right)}{f_2\left(\overline{A}_{r/2}\right)} c_1^2.$$

Fix some $\rho < r/16$ and assume that there is a constant $c_3 > 0$ such that

$$\frac{f_k(x,t)}{f_{3-k}(x,t)} \geq \frac{f_k(y,u)}{f_{3-k}(y,u)} c_3 \tag{6.11}$$

for $k = 1, 2$, $(x,t), (y,u) \in \Psi_\rho$. Let $\mu_{(x,t)}$ denote the caloric measure on $\partial\Psi_\rho$ for $(x,t) \in \overline{\Psi_\rho}$ and $\underline{\Delta}_\rho \stackrel{\mathrm{df}}{=} \{(x,t) \in \partial\Psi_\rho : t = s - \rho^2\}$. Recall that $\mu_{(x,t)}(\,\cdot\,)$ does not charge $\{(y,u) \in \partial\Psi_\rho : u = s + \rho^2\}$. For a Borel measurable set $B \subset \underline{\Delta}_\rho$, the function $(x,t) \to \mu_{(x,t)}(B)$ is parabolic in Ψ_ρ and vanishes continuously on

$$\left\{(x,t) \in \partial\Psi_\rho : t \in \left(s - \rho^2, s + \rho^2\right)\right\}.$$

It follows from Corollary 2.2 of Fabes et al. (1986) that

$$\mu_{\underline{A}_{\rho/8}}(B) \geq c_4 \mu_{\overline{A}_{\rho/8}}(B). \tag{6.12}$$

The constant c_4 depends only on n and λ although in the paper of Fabes et al. (1986) it depends on the diameter of Ψ_ρ as well. The last dependence may be removed by scaling.

Let B, C be Borel measurable sets in $\underline{\Delta}_\rho$. If $\mu_\cdot(B) > 0$ and $\mu_\cdot(C) > 0$ on Ψ_ρ, then by Theorem 1.6 of Fabes et al. (1986) and (6.12) we have

$$\begin{aligned}\frac{\mu_{(x^1,t^1)}(B)}{\mu_{(x^1,t^1)}(C)} &\geq c_1 \frac{\mu_{\underline{A}_{\rho/8}}(B)}{\mu_{\overline{A}_{\rho/8}}(C)} \\ &\geq c_1 c_4^2 \frac{\mu_{\overline{A}_{\rho/8}}(B)}{\mu_{\underline{A}_{\rho/8}}(C)} \\ &\geq c_1^2 c_4^2 \frac{\mu_{(x^2,t^2)}(B)}{\mu_{(x^2,t^2)}(C)}\end{aligned} \tag{6.13}$$

for $(x^1, t^1), (x^2, t^2) \in \Psi_{\rho/64}$. By the forward and backward Harnack principles (see Theorem 0.2 and Theorem 2.1 of Fabes et al. (1986)), if $\mu_{(x,t)}(B) = 0$ for some $(x,t) \in \Psi_\rho$, then $\mu_\cdot(B) \equiv 0$ in Ψ_ρ, and similarly for $\mu_\cdot(C)$. Thus it follows by our convention for zero divisors that (6.13) holds for all Borel measurable sets $B, C \subset \underline{\Delta}_\rho$. Fix some $(x^0, t^0) \in \partial\Psi_\rho$ such that $t^0 = s + \rho^2, |x^0| < \rho, x^0 \in D$. Then for each $(x,t) \in \Psi_\rho$, the caloric measure $\mu_{(x,t)}$ is absolutely continuous with respect to $\mu_{(x^0,t^0)}$ (Fabes et al. (1986), page 540). Let $g_{(x,t)}$ denote the Radon-Nikodym derivative $d\mu_{(x,t)}/\,d\mu_{(x^0,t^0)}$ on $\partial\Psi_\rho$. Then (6.13) implies that

$$\frac{g_{(x^1,t^1)}\,(y^1,u^1)}{g_{(x^1,t^1)}\,(y^2,u^2)} \geq c_1^2 c_4^2 \frac{g_{(x^2,t^2)}\,(y^1,u^1)}{g_{(x^2,t^2)}\,(y^2,u^2)} \tag{6.14}$$

for $(x^k, t^k) \in \Psi_{\rho/64}, (y^k, u^k) \in \underline{\Delta}_\rho, k = 1, 2$. As above, we assume our convention about zero divisors here. Although strictly speaking (6.14) only holds for $\mu_{(x^0,t^0)}$-a.e. $(y^k, u^k) \in \underline{\Delta}_\rho$, by changing $g_{(x^1,t^1)}$ and $g_{(x^2,t^2)}$ on a set of $\mu_{(x^0,t^0)}$-measure zero (possibly depending on (x^1,t^1), (x^2,t^2)), we can make (6.14) hold for all (y^k, u^k), $k = 1, 2$, as indicated.

Fix a point $y^0 \in D$ with $|y^0| > r$ and let $G_D(\cdot\,,\cdot)$ be the Green function in D. Then $G(x,t) \stackrel{\text{df}}{=} G_D(y^0, x)$ is parabolic in Ψ_r and vanishes on Δ_r. Choose $c_5 = c_5(n, \lambda) > 0$ so small that the ball B_1 with center $(\widetilde{0}, \rho/2)$ and radius $2\rho c_5$ is contained in D. Let B_2 be the concentric ball with half the radius of B_1. By the elliptic Harnack principle, $G(x,t) \geq c_6 G(y,u)$ for $x, y \in B_2,\ t, u > 0$ and $c_6 = c_6(n)$.

Apply Theorem 1.6 of Fabes et al. (1986) to see that

$$\frac{f_k(x^1,t^1)}{G(x^1,t^1)} \geq c_1 \frac{f_k(\underline{A}_{r/2})}{G(\overline{A}_{r/2})},$$
$$\frac{G(x^2,t^2)}{f_k(x^2,t^2)} \geq c_1 \frac{G(\underline{A}_{r/2})}{f_k(\overline{A}_{r/2})},$$

and, therefore,

$$\begin{aligned} \frac{f_k(x^1,t^1)}{f_k(x^2,t^2)} &\geq c_1^2 \frac{f_k(\underline{A}_{r/2})}{f_k(\overline{A}_{r/2})} \frac{G(\underline{A}_{r/2})}{G(\overline{A}_{r/2})} \frac{G(x^1,t^1)}{G(x^2,t^2)} \\ &\geq c_1^2 c_6^2 \frac{f_k(\underline{A}_{r/2})}{f_k(\overline{A}_{r/2})} \end{aligned} \tag{6.15}$$

for $(x^1,t^1), (x^2,t^2) \in \Psi_{r/16}, x^1, x^2 \in B_2$.

For $k = 1, 2, (x,t) \in \Psi_\rho$, let

$$\widetilde{f}_k(x,t) = \int_{\underline{\Delta}_\rho} f_k(y,u)\,d\mu_{(x,t)}(y,u). \tag{6.16}$$

The function $\widetilde{f}_k$ is parabolic in Ψ_ρ and vanishes continuously on Δ_ρ, since f_k vanishes on $\Delta_\rho \subset \Delta_r$. Let $\underline{B}_2 = \{(x,t) \in \underline{\Delta}_\rho : x \in B_2\}$. It is easy to see that

$$\mu_{\underline{A}_{\rho/2}}(\underline{B}_2) \geq c_7 = c_7(n,\lambda) > 0. \tag{6.17}$$

By (6.15),

$$f_k(y,u) \geq f_k(\overline{A}_{\rho/2}) c_1^2 c_6^2 \frac{f_k(\underline{A}_{r/2})}{f_k(\overline{A}_{r/2})} \tag{6.18}$$

for $(y,u) \in \underline{B}_2, k = 1, 2$. Combine (6.16), (6.17) and (6.18) to see that

$$\widetilde{f}_k(\underline{A}_{\rho/2}) \geq c_7 f_k(\overline{A}_{\rho/2}) c_1^2 c_6^2 \frac{f_k(\underline{A}_{r/2})}{f_k(\overline{A}_{r/2})}.$$

Theorem 1.6 of Fabes et al. (1986) implies that

$$\begin{aligned} \frac{\widetilde{f}_k(x,t)}{f_k(x,t)} &\geq c_1 \frac{\widetilde{f}_k(\underline{A}_{\rho/2})}{f_k(\overline{A}_{\rho/2})} \\ &\geq c_1^3 c_6^2 c_7 \frac{f_k(\underline{A}_{r/2})}{f_k(\overline{A}_{r/2})} \end{aligned} \tag{6.19}$$

for $(x,t) \in \Psi_{\rho/16}, k = 1, 2$.

Now Lemma 6.1 will be applied with $f_1(v), f_2(v), g_x(v), g_y(v), W$ and V replaced by $f_1(z,v), f_2(z,v), g_{(x,t)}(z,v), g_{(y,u)}(z,v), \partial\Psi_\rho$ and $\underline{\Delta}_\rho$. Note that

$$f_k(x,t) = \int_{\partial\Psi_\rho} f_k(z,v)\,d\mu_{(x,t)}(z,v) = \int_{\partial\Psi_\rho} f_k(z,v) g_{(x,t)}(z,v)\,d\mu_{(x_0,t_0)}(z,v),$$

for $k=1,2,(x,t)\in\Psi_\rho$. Let

$$c_8=c_1^3c_6^2c_7\min\left(\frac{f_1(\underline{A}_{r/2})}{f_1(\overline{A}_{r/2})},\frac{f_2(\underline{A}_{r/2})}{f_2(\overline{A}_{r/2})}\right).$$

Observe that (6.11) extends to $(x,t),(y,u)\in\partial\Psi_\rho$ by the continuity of $f_k, k=1,2$, and our convention about zero divisors. With this, (6.14) and (6.19), the hypotheses of Lemma 6.1 are verified and so

$$\frac{f_k(x,t)}{f_{3-k}(x,t)}\geq\frac{f_k(y,u)}{f_{3-k}(y,u)}c_9$$

for $(x,t),(y,u)\in\Psi_{\rho/64}$, $k=1,2$ and $c_9=c_3+(1-c_3)\,c_1^4c_4^4c_8^2$.

It then follows by induction from the above that

$$\frac{f_k(x,t)}{f_{3-k}(x,t)}\geq\frac{f_k(y,u)}{f_{3-k}(y,u)}c_{10}$$

for $k=1,2,(x,t),(y,u)\in\Psi_{r\cdot2^{-6m}}$ and

$$c_{10}=c_{10}(m),$$

$$c_{10}(1)=c_2,$$

$$c_{10}(m+1)=c_{10}(m)+(1-c_{10}(m))\,c_1^4c_4^4c_8^2,\quad m\geq1.$$

Note that $c_1,c_4,c_8\in(0,1)$, so $c_{10}(m)$ is increasing as $m\to\infty$ and, moreover, $c_{10}(m)\to1$. It is easy to check that c_{10} depends only on m,n,λ, $f_1(\underline{A}_{r/2})/f_1(\overline{A}_{r/2})$ and $f_2(\underline{A}_{r/2})/f_2(\overline{A}_{r/2})$ and it is an increasing function of $f_1(\underline{A}_{r/2})/f_1(\overline{A}_{r/2})$ and $f_2(\underline{A}_{r/2})/f_2(\overline{A}_{r/2})$. ∎

Suppose that for $k=1,2,\varphi^k$ is a Lipschitz function with constant $\lambda,\varphi^k:\mathbf{R}^{n-1}\to\mathbf{R},\varphi^k(0)=0,D^k=\{x\in\mathbf{R}^n:x_n>\varphi^k(\widetilde{x})\}$ and let $\Psi_r^k(x,t),\Delta_r^k(x,t)$ etc. be defined relative to D^k.

COROLLARY 6.1. *There exists $c = c(a_1, a_2, a_3, a_4, n, \lambda, \varepsilon)$ with the following properties.*

(i) $c \in (0,1)$, c is increasing in ε and decreasing in a_1, a_2, a_3 and a_4.

(ii) For fixed a_1, a_2, a_3, a_4, n and λ,

$$\lim_{\varepsilon \downarrow 0} c(a_1, a_2, a_3, a_4, n, \lambda, \varepsilon) = 0.$$

(iii) Suppose that for $k = 1, 2$, functions f_k and g_k are positive and parabolic in $\Psi_r^k(0,s)$ and vanish continuously on $\Delta_r^k(0,s)$, where $s > 0$, $0 < r < \sqrt{s}$. Assume that

$$\frac{f_1(x^1,t^1)}{f_2(x^1,t^1)} \frac{g_2(x^1,t^1)}{g_1(x^1,t^1)} \in (1-\varepsilon, 1+\varepsilon) \tag{6.20}$$

for some $(x^1,t^1) \in \Psi_\varepsilon^1(0,s) \cap \Psi_\varepsilon^2(0,s)$, $\varepsilon < r/16$. Then

$$\lim_{\substack{(x,t)\to(0,s)\\(x,t)\in\Psi_r^1(0,s)}} \frac{f_1(x,t)}{g_1(x,t)} \cdot \lim_{\substack{(y,u)\to(0,s)\\(y,u)\in\Psi_r^2(0,s)}} \frac{g_2(y,u)}{f_2(y,u)} \in (1-c, 1+c) \tag{6.21}$$

where

$$c = c\left(\frac{f_1(\underline{A}_{r/2}(0,s))}{f_1(\overline{A}_{r/2}(0,s))}, \frac{f_2(\underline{A}_{r/2}(0,s))}{f_2(\overline{A}_{r/2}(0,s))}, \frac{g_1(\underline{A}_{r/2}(0,s))}{g_1(\overline{A}_{r/2}(0,s))}, \frac{g_2(\underline{A}_{r/2}(0,s))}{g_2(\overline{A}_{r/2}(0,s))}, n, \lambda, \varepsilon\right).$$

In particular, the limits in (6.21) exist.

Proof: Let c_1 denote the constant c in Theorem 6.1 (iii) with f_1, g_1 in place of f_1, f_2 there. Then,

$$\frac{g_1(x^1,t^1)}{f_1(x^1,t^1)} \frac{f_1(x,t)}{g_1(x,t)} \in (c_1, c_1^{-1}) \tag{6.22}$$

for $(x,t) \in \Psi_\varepsilon^1(0,s)$. Similarly, let c_2 denote the constant c obtained in Theorem 6.1 (iii) with f_2, g_2 in place of f_1, g_1. Then

$$\frac{f_2(x^1,t^1)}{g_2(x^1,t^1)} \frac{g_2(y,u)}{f_2(y,u)} \in (c_2, c_2^{-1}), \tag{6.23}$$

for $(y,u) \in \Psi_\varepsilon^2(0,s)$. By multiplying (6.22) and (6.23) and using (6.20) we obtain

$$\frac{f_1(x,t)}{g_1(x,t)}\frac{g_2(y,u)}{f_2(y,u)} \in (c_1c_2(1-\varepsilon), c_1^{-1}c_2^{-1}(1+\varepsilon)) \tag{6.24}$$

for $(x,t) \in \Psi_\varepsilon^1(0,s)$ and $(y,u) \in \Psi_\varepsilon^2(0,s)$. The existence of the limits in (6.21) follows immediately from Theorem 6.1. The existence of a c such that (i)-(iii) hold then follows from (6.24) and the properties of c_1, c_2. ∎

REFERENCES

[1] Bañuelos, R. and Burdzy, K., *A representation of the local time on Lipschitz surfaces*, (forthcoming paper).

[2] Bass, R., *Joint continuity and representations of additive functionals of d-dimensional Brownian motion*, Stoch. Proc. Appl. **17** (1984), 211-227.

[3] Brémaud, P., "Point Processes and Queues. Martingale Dynamics," Springer, New York, 1981.

[4] Burdzy, K., "Multidimensional Brownian Excursions and Potential Theory," Longman, London, 1987.

[5] Burdzy, K. and Williams, R. J., *On Brownian excursions in Lipschitz domains; Part I. Local path properties*, Trans. Amer. Math. Soc. **298** (1986), 289-306.

[6] Doob, J. L., "Classical Potential Theory and Its Probabilistic Counterpart," Springer, New York, 1984.

[7] Fabes, E. B., Garofalo, N. and Salsa, S., *A backward Harnack inequality and Fatou theorem for nonnegative solutions of parabolic equations*, Illinois J. Math. **30** (1986), 536-565.

[8] Fristedt, B. and Taylor, S.J., *Constructions of local time for a Markov process*, Z. Wahrscheinlichkeitstheorie verw. Gebiete **62** (1983), 73-112.

[9] HSU, P., *On excursions of reflecting Brownian motion*, Trans. Amer. Math. Soc. **296** (1986), 239-264.

[10] MAISONNEUVE, B., *Exit systems*, Ann. Probab. **3** (1975), 399-411.

[11] REVUZ, D., *Mesures associees aux fonctionelles additives de Markov, I.*, Trans. Amer. Math. Soc. **148** (1970), 501-531.

[12] WILLIAMS, D., "Diffusions, Markov Processes and Martingales, I," Wiley, New York, 1979.

Krzysztof Burdzy
Department of Mathematics
University of Washington
Seattle, WA 98195

Ellen H. Toby
Department of Mathematics
and Computer Sciences
University of California
Riverside, CA 92521

Ruth J. Williams
Department of Mathematics
University of California
San Diego, CA 92093

GAUGE THEOREM FOR UNBOUNDED DOMAINS

by

Kai Lai Chung*

Let $\{X_t, t>0\}$ be the Brownian motion process in R^d, $d>1$; D a domain (nonempty, open and connected set) in R^d; q a Borel function on D. Put

$$\tau_D = \inf\{t>0: X_t \notin D\},$$

and

(1) $$u(x) = E^x\{\tau_D<\infty;\ \exp[\int_0^{\tau_D} q(X_t)dt]\}$$

where E^x (P^x) denotes the expectation (probability) under $X_0 = x$. The function u is called the gauge for (D,q), provided it is well-defined, namely when the integral involved exists. A result of the following form is called gauge theorem:

(2) either $u \equiv +\infty$ in D, or u is bounded in D.

Let $\overline{D}$ denote the closure of D in R^d (no point at infinity). It is easy to show that if it is bounded in D, then the same upper bound serves for u in $\overline{D}$, so

*Research supported in part by AFOSR Grant 85-0330.

that u is in fact bounded in R^d since it is equal to one in $R^d - \overline{D}$. In this case we say that (D,q) is gaugeable.

The gauge theorem was first established by Chung and Rao [1] when $m(D) < \infty$ where m denotes the Lebesque measure and $q \in L^\infty(D)$. Subsequently a large class of q was studied by Aizenman and Simon [2], which is known as the Stummel-Kato class J_d. This class of functions is characterized by the following condition:

$$(3) \qquad \lim_{\alpha \downarrow 0} \sup_{x \in R^d} \int_{|x-y|<\alpha} K_d(x-y)|q(y)|dy = 0,$$

where for $|u| < 1$:

$$K_d(u) = \begin{cases} |u|, & \text{if } d = 1; \\ -\log|u|, & \text{if } d = 2; \\ |u|^{2-d} & \text{if } d \geq 3. \end{cases}$$

If q is defined on R^d, and for each bounded domain D we have $1_D q \in K_d$, then we denote this by $q \in K_d^{loc}$. The gauge theorem for bounded D and $q \in K_d^{loc}$ was proved by Z. Zhao [3].

Recently Zhao and I observed tha there is an extension of the gauge theorem to unbounded domains, as follows.

THEOREM 1. The gauge theorem holds for any domain D in R^d, $d \geq 3$, provided that

$$(4) \qquad 1_D q \in K_d \cap L^1(D).$$

Since it is known that $q \in K_d$ implies $q \in L^1(D)$ for bounded D, this includes Zhao's result in case $d \geq 3$. The proof of Theorem 1 turns out to be quite easy owing to the fact that for $d \geq 3$, the Green function G_D for any D is dominated by a constant multiple of $K_d(x-y) = |x-y|^{2-d}$, so that

$$(5) \qquad E^x\{\int_0^{\tau_D} |q(X_t)|dt\} = G_D|q|(x) = \int_{R^d} G_D(x,y)|q(y)|dy$$

is bounded by (3) and (4). The old method of proof of Theorem 1 when D is bounded then carries over to the general case without difficulty.

For $d = 1$, the followng example by K. B. Erickson shows that the analogue of Theorem 1 is false:

$$D = (1,\infty), \quad q(x) = \frac{c}{x^2}, \quad 0 < c < 1/8;$$

$$u(x) = \chi^a \quad \text{where} \quad a = \frac{1 - \sqrt{1-8c}}{2}.$$

More recently, V. Papanicolaou gave a counterexample to the analogue of Theorem 1 for $d = 2$. Roughly stated, the example above is transformed radially into a planar domain D which is the complement of a disc, and q is not only integrable over D but even bounded in D.

There remains the question whether the gauge theorem holds for a domain D in R^2 with $m(D) < \infty$, and $q \in K_2$. This is an extension of the original result by Chung and Rao. Now the precise definition of Green's function for an unbounded D in R^2 is not easy to find in classical texts of analysis. By contrast, the following probabilistic definition holds in all dimensions and all domains:

$$G_D(x,y) = \int_0^\infty p^D(t;x,y)dt \tag{6}$$

where p^D is the transition probability density of the Brownian motion killed outside D; see Hunt [4]. Observe that no regularity of ∂D is assumed. When $d = 2$ and D is arbitrary, the quantity in (6) may be identically $+\infty$ in $D \times D$. The domain D is called "Greenian" when G_D is finite. But even for such D the boundedness of G_D is not a simple question. It follows from (6) that $G_D = 0$ outside $\overline{D} \times \overline{D}$; a deeper result implies that $G_D = 0$ almost everywhere on $\partial D \times \partial D$. It is customary to put $G = 0$ outside $D \times D$, where $G_D(x,y)dy$ or $G_D(x,y)dx$ is in question owing to the last remark. Thus it is customary to replace the integral over R^d in (5) by the integral over D. When D is unbounded, it is possible that $q \in K_2$ but $q \notin L^1(D)$. A simple example was given by V. Papanicolaou. Hence even if G_D is bounded in such a D, it does not follow that the integral in (5) must be finite. These problems are mentioned here in order to underscore the difficulty of dealing with the analytical formula for $G_D|q|$, in comparison with the situation in R^d, $d > 3$. Specifically, the following analytic problem still awaits an <u>analytic</u> solution.

Problem. Let D be a domain in R^2 with $m(D) < \infty$, G_D the Green function for D, and $1_D q \in K_2$. Show that

$$\int_D G_D(x,y)q(y)dy$$

is bounded.

The solution of this problem by probabilistic methods will now be given. The gauge theorem in the missing case will then be proved by a strengthening of the result, together with the general arguments in previous work which are readily applicable.

The probabilistic approach ignores the Green function and begins with the first expression given in (5), which is the proper setting. Next, instead of the analytic definition of K_d which proves to be inconvenient in the present case, an equivalent probabilistic characterization due to Aizenman and Simon [2] will be employed. According to the latter, $q \in K_d$ if and only if

$$\lim_{t \downarrow 0} \sup_{x \in R^d} E^x \{\int_0^t |q(X_s)| ds\} = 0. \tag{7}$$

This is another instance where the probabilistic formulation is so much neater than the analytic one. There is an immediate application which has been overlooked, namely we can now affirm that the gauge u is well defined when $q \in K_d$. To see this let us put

$$\phi(t) = \sup_{x \in R^d} E^x\{\int_0^t |q(X_s)| ds\}.$$

Then it follows from the Markov property of $\{X_t\}$ that ϕ is subadditive, i.e.

(8) $$\phi(t+t') < \phi(t) + \phi(t')$$

for $t > 0$, $t' > 0$. Since $\lim_{t\downarrow 0} \phi(t) = 0$ by (7), it is trivial by (8) that $\phi(t) < \infty$ far all $t < \infty$, and consequently almost surely:

$$\int_0^t |q(X_s)|ds < \infty.$$

Therefore the same is true when the t above is replaced by any finite random variable. In particular

$$\int_0^{\tau_D} |q(X_s)|ds < \infty \quad \text{on} \quad \{\tau_D < \infty\}.$$

It follows that the gauge u in (1) is not only well-defined for $q \in K_d$, but indeed $0 < u < +\infty$. This result was proved in the past by an unnecessary recourse to Jensen's inequality, and only under extraneous assumptions. The next lemma contains the solution of the Problem posed above.

THEOREM 2. Let D be a domain in R^d, $d > 1$, with $m(D) < \infty$; and $1_D q \in K_d$. Then

(i) $G_D|q|$ is bounded in R^d;

(ii) if $m(D) \to 0$, then $G_D|q|$ converges to zero uniformly in R^d.

PROOF. We recall first that there exists a constant C_d depending only on d such that for all $x \in R^d$:

$$(9) \qquad E^x\{\tau_D\} \leq C_d m(D)^{2/d}$$

It follows that $m(D) < \infty$ implies $\tau_D < \infty$ a.s. and we may omit "$\tau_D < \infty$" in (1). Next we recall a result due to Khasminskii: for integer $k \geq 2$,

$$(10) \qquad \sup_{x\in R^d} E^x\{[\int_0^T |q(X_s)|ds]^k\} \leq k! \sup_{x\in R^d} E^x\{\int_0^T |q(X_s)|ds\}^k$$

where T is either a constant t or τ_D.

Now let $\varepsilon > 0$ be given, then by (7) there exists $\delta > 0$ such that $\phi(\delta) < \varepsilon$. We have for all x:

$$(11) \qquad \begin{aligned} E^x\{\int_0^{\tau_D}|q(X_s)|ds\} &\leq E^x\{\int_0^{\delta} |q(X_s)|ds\} \\ &+ \sum_{n=1}^{\infty} E^x\{n\delta<\tau_D \leq (n+1)\delta; \int_0^{(n+1)\delta} |q(X_s)|ds\} \end{aligned}$$

where "$\tau_D < \infty$" has been importantly used. The first term on the right side above is less than ε. The general term in the infinite series is bounded by

$$(12) \qquad P^x\{n\delta < \tau_D\}^{1/2}\cdot \quad E^x\{[\int_0^{(n+1)\delta} |q(X_s)|ds]^2\}^{1/2}$$

by Cauchy-Schwarz inequality followed by an application of (10) with $T = \tau_D$ and $k = 2$. It follows from (8) that

$$(13) \qquad E^x\{\int_0^{(n+1)\delta} |q(X_s)|ds\} \leq (n+1)\varepsilon.$$

Next, as a particular case of (10) with $q \equiv 1$ and $T = \tau_D$, we have for positive integer k:

(14) $$\sup_{x\in R^d} E^x\{\tau_D^k\} \le k! \sup_{x\in R^d} E^x\{\tau_D\}^k.$$

Hence by Chebyshev's inequality followed by (9), for all x:

(15) $$P^x\{t < \tau_D\} \le k!(\frac{C_d}{t})^k \, m(D)^{2k/d}$$

Taking $k = 5$ and merging constants, we obtain the following upper bouond for the quantity in (12):

$$C \frac{1}{(n\delta)^{5/2}} m(D)^{5/d}(n+1)\varepsilon$$

where C is a constant depending only on d. Summing over $n \ge 1$, we conclude from (11) that

(16) $$\sup_{x\in R^d} G_D|q|(x) \le \varepsilon + \frac{C'\varepsilon}{\delta^{5/2}} m(D)^{5/d}$$

where C' is another constant like C, and $\delta = \delta(\varepsilon)$.

Assertion (i) follows from (16) with $\varepsilon = 1$, say. Furthermore, if

$$m(D) \le (C')^{-d/5}\delta(\varepsilon)^{d/2}$$

then the right member of (16) is less than 2ε. This is the meaning of assertion (ii).

Remark. Part (i) of the theorem was proved independently by Papanicolaou.

To proceed to the gauge theorem, we begin by stating the following result known as Harnack's inequality. This was first proved in [1] for $q \in L^\infty(D)$, extended to

$q \in K_d^{loc}$ in [2], and simplified in [3].

THEOREM 3. Let D be an arbitrary domain in R^d, $d \geq 1$, and $q \in K_d^{loc}$. If there exists $x_0 \in D$ such that $u(x_0) < \infty$, then for any compact subset C of D which contains x_0 there exists a constant A depending only on D, C and q such that

$$\sup_{x \in C} u(x) \leq Au(x_0).$$

THEOREM 4 (Gauge Theorem). Let D be a domain in R^d, $d \geq 1$, with $m(D) < \infty$; and $1_D q \in K_d$. If there exists $x_0 \in D$ such that $u(x_0) < \infty$, then u is bounded in $\overline{D}$.

PROOF. Under the assumption there exists a compact subset of C of D, with $x_0 \in C$ such that $D-C$ is connected and $m(D-C)$ so small that

$$\sup_{x \in D-C} G_{D-C}|q|(x) < \varepsilon < 1. \tag{17}$$

This is possible by assertion (ii) of Theorem 2. It follows from Khasminskii's lemma (see [2]), indeed as a straightforward consequence of (10) that

$$\sup_{x \in D-C} E^x\{\exp[\int_0^{\tau_{D-C}} |q(X_t)|dt]\} < \frac{1}{1-\varepsilon} < \infty. \tag{18}$$

Thus $(D-C, q)$ is gaugeable. Next by Theorem 3, there is a constant M such that

(19) $$\sup_{x \in C} u(x) = M < \infty.$$

Now the original proof of the gauge theorem given as Theorem 1.2 in [1], with $f \equiv 1$ and $E = D-C$, goes through without change, if we use (7) here instead of the old (16) there. The result is

$$\sup_{x \in \bar{D}} u(x) < \frac{1}{1-\varepsilon}(M+1).$$

Once the gauge theorm is established, the rest of the development in [1] unfolds in the same way. For instance, the continuity of u in D is proved as follows. Let $x \in D$, $B = B(x)$, a disk or ball with x as center, and with closure in D. Then by the strong Markov property u has the following representation in B:

(2) $$u(x) = E^x\{\exp[\int_0^{\tau_B} q(X_s)ds].\ u(X(\tau_B))]\}, \quad x \in B.$$

It is known that the gaugeability of (D,q) implies that of (B,q). As u is bounded on ∂B by the gauge theorem the right member of (20) is known to be a (weak) solution of the Schröedinger equation in B:

(21) $$(\frac{\Delta}{2} + q)u = 0.$$

The continuity of u in B is part of this proposition. Since x is arbitrary, u is continuous in D, and is indeed a solution of the equation in D. It is further proved in [1] that at each regular boundary point $z \in \partial D$, we have

$$(22) \qquad \lim_{\overline{D} \ni x \to z} u(x) = 1.$$

Let us summarize these properties of the gauge, together with an earlier remark, in the final theorem below.

THEOREM 5. If the gauge is not identically $+\infty$ in D, then it is strictly positive, bounded and continuous in $\overline{D}$. Moreover, it is a solution of (21) with boundary value 1 at all regular point of ∂D.

Remember that when D is unbounded, "∞" is not regarded as a boundary point. Indeed, the behavior of $u(x)$ as $|x| \to \infty$ is not mentioned above. In R^1, simple examples show that u may not have a limit at ∞, even when it is bounded. Presumably this is the case in all dimensions.

References

[1] K. L. Chung and K. M. Rao, Feynman-Kac functional and the Schrödinger equation, Seminar in Stochastic Processes 1(1981), 1-29.

[2] M. Aizenman and B. Simon, Brownian motion and Harnack inequality for Schrödinger operators, Comm. Pure Appl. Math. 35(1982), 209-273.

[3] Z. Zhao, Conditional gauge with unbounded potential, Z. Wahrscheinlichkeitstheorie Verw. Geb. 65(1983), 13-18.

[4] G. A. Hunt, Some theorems concerning Brownian motion, Trans. Amer. Math. Soc. 81(1956), 294-319.

Reminiscences of some of Paul Lévy's ideas in Brownian Motion and in Markov Chains*

KAI LAI CHUNG

We begin with a resume. Let $\{P(t),\ t \geq 0\}$ be a semigroup of stochastic matrices with elements $p_{ij}(t), (i,j) \in I \times I$, where I is a countable set, satisfying the condition

$$\lim_{t \downarrow 0} p_{ii}(t) = 1. \tag{1}$$

It is known that $p'_{ij}(0) = q_{ij}$ exists and

$$0 \leq q_i = -q_{ii} \leq +\infty, \quad 0 \leq q_{ij} < \infty,\ i \neq j; \tag{2}$$

$$\sum_{j \neq i} q_{ij} \leq q_i. \tag{3}$$

The state i is called *stable* if $q_i < +\infty$, and *instantaneous* if $q_i = +\infty$ (Lévy's terminology). The matrix $Q = (q_{ij})$ is called *conservative* when equality holds in (3) for all i.

If the convergence in (1) holds uniformly with respect to all i, or equivalently if the set of all q_i is bounded, then we have

$$P(t) = e^{Qt} \tag{4}$$

Let $\{X(t),\ t \geq 0\}$ be a Markov chain with $P(t)$ as its transition matrix, separable and measurable. Then in the special case just mentioned, almost all sample functions are step functions in any finite time interval. The Poisson process is an example, as well as the case when I is a finite set.

Before Lévy, the regularity properties of the sample functions of a general Markov chain have been investigated by Doob by martingale methods (1942, 1945). To describe the *allure* of a typical path, let us start it at a stable state i. The Markov property implies that it will remain at i during a sojourn time ρ_1 with $P(\rho_1 > t) = e^{-q_i t}$. Unless $q_i = 0$, ρ_1 is finite but $X(\rho_1+)$ need not exist if inequality holds in (3), or if there is some instantaneous j; in fact the path may encounter an infinity of states immediately after ρ_1 and so the analysis is halted. To avoid such a quick termination let us assume that all states are stable and the Q-matrix is conservative, also that all $q_i > 0$ to exclude a trivial case. Then at the time ρ_1 the path will jump from i to j with probability q_{ij}/q_i for all $j \neq i$, and we can resume

* Address delivered on June 22, 1987 at the Colloque Paul Lévy, Palaiseau, France.

the analysis starting with j. The path will remain in j during another sojourn time ρ_2 with $P(\rho_2 > t) = e^{-q_j t}$, then jump again, and so on. The analysis proceeds by induction until the time

$$\rho_1 + \rho_2 + \ldots + \rho_n + \ldots = \tau.$$

If $\tau = +\infty$ then the entire path has been traced, as in the Poisson case. The discovery of the possibility that τ may be finite with positive or even full probability caused a sensation, and much confusion. Read the Prologue of my Strasbourg Lectures for some historical perspective. The tracing of the path has been stopped in its track, what happens after τ? Confusion arose because a wrong question: "what can we do after τ?" was asked. The proper question is of course "what will the path do after τ?". The path exists, and there is nothing we can do except to find it! Looking back, we now realize that the problem lies in the insufficiency of the initial date given by the Q-matrix, and further structure of the paths must be searched out. This leads to a boundary theory which yields new clues to the paths but cannot deal with the general situation. I do not believe there is any complete solution, and certainly bulldozing the countable state space into something unrecognizable is no solution at all.

Lévy forsook the old way of tracking the path and instead plunged in midstream, as it were. If i is stable it is intuitively clear that the set $\{t : X(t) = i\}$ is a collection of disjoint (maximal) intervals. He made the crucial observation that the number of these i-intervals is finite up to any finite time (almost surely). This requires a proof; a short one is given in my Strasbourg Lectures. Once this is established, it follows that the i-intervals may be ordered in sequence and that their lengths form independent and identically distributed random variables. Moreover, these lengths are also independent of "everything outside the intervals". Hence if all states are stable, this global picture gives a bird's-eye view of the paths, with an abundance of mutual independence among various portions thereof. He applied this idea to the following theorem, one of the finest in the theory. (In his 1951 paper he attributed its origin to this result, which I discussed with him at the Berkeley Symposium a year before. Thus some conferences yield fruits.)

Theorem. *For any i and j, either $p_{ij}(t) = 0$ for all $t \geq 0$, or $p_{ij}(t) > 0$ for all $t > 0$.*

The case $i = j$ is easy. Now assume all states stable and $i \neq j$. Suppose $p_{ij}(t_0) > 0$, then $P_i\{T_j \leq t_0\} > 0$, where P_i is the probability starting from i, and T_j is the hitting probability of j. Let $I_n(k)$ denote the nth k-interval, and $|I_n(k)|$ its length. Then the global description above implies that

$$T_j = \Sigma |I_n(k)|$$

where the sum is taken over all $I_n(k)$ contained in $[0, T_j)$. By hypothesis $T_j < \infty$ on a set of positive probability. Hence by Egorov's theorem, for any $\epsilon > 0$ there exists finite integers K and N such that

$$T_j = S + R \leq S + \frac{\epsilon}{2},$$

where S is the sum $\sum$ restricted to $k \leq K$ and $n \leq N$, and R is the rest. There are only a finite number of permutations of the $I_n(k)$'s after the restriction, hence there is a subset $\wedge$ of the previous set with $P(\wedge) > 0$ on which

$$T_j = \sum_{v=1}^{m} |I_{n_v}(k_v)| + R, \text{ and } R \leq \frac{\epsilon}{2}, \tag{5}$$

where $\{n_v, k_v, 1 \leq v \leq m\}$, $k_v \leq K, n_v \leq N$, is a fixed sequence not depending on the sample function. The set $\wedge$ is defined by a specific ordering of the intervals, hence it is independent of their lengths. It follows that

$$P(\wedge; \tau \leq \epsilon) \geq P\left(\wedge; R \leq \frac{\epsilon}{2}\right) P\left(\sum_{v=1}^{m} |I_{n_v}(k_v)| \leq \frac{\epsilon}{2}\right) > 0$$

because each $|I_n(k)|$ is exponentially distributed. Q. E. D.

This proof does not seem to extend to the case when there are instantaneous states. D. G. Austin first proved the general case by a brilliant probabilistic argument using the right separability of the process and Lebesgue's theorem on differentiation of monotone functions. Later D. Ornstein gave another more analytic proof. All three proofs are given in my book on Markov chains. An exposition of Lévy's proof was included in R. V. Chacon's dissertation as a special assignment.

Lévy gave a tantalizing example of a Markov chain with only stable states and no jumps at all, all discontinuities being of the second kind; in particular all $q_{ij} = 0$ for $i \neq j$. Take a strictly increasing function on $[0, \infty)$ with jumps at all the rationals, the size of the jump J_r at r being randomized with $P(J_r > t) = e^{-q_r t}$, and all J_r's are independent. The set of q_r's are chosen as follows:

$$\forall r: \; q_{r+1} = q_r; \; \sum_{0 \leq r < 1} (1/q_r) < \infty.$$

This will ensure that almost all the functions are finite and increases to infinity. The right continuous inverse of each such function is a continuous singular monotone function. There is a Markov chain whose sample functions are the collection of these inverses. Thus each path goes through all the positive reals in their natural order, sojourning in each rational but passing through all the irrationals in (Lebesgue) null time. This example is a veritable Columbus's egg stood on its flattened head. It is possible to write down explicit formulas for $X(t)$ as well as $P(t)$, as I did in my book, but it is a tedious and not very enlightening task. Lévy gave a number of such examples of Markov chains by prescribing the sample functions. Often they seem intuitively clear but require painstaking verification *après coup*. This provokes a curious, and I think important question: are there more effective ways of recognizing a Markov process without going through the usual formalities?

A week ago I received the fourth instalment of Dellacherie-Meyer's tomes, which gives an account of Lévy's ideas on local time and excursions of a Brownian motion on the line, together with later developments. Time being short I may therefore confine myself to a few remarks to fill some gaps, perhaps.

(I) Lévy derived a multitude of formulas for the excursions by means of the equivalent process:

$$Y_t = M_t - X_t,$$

where X is the Brownian motion, and $M_t = \max_{0 \le s \le t} X_s$. He showed that Y and $|X|$ are equivalent processes, so their zero-sets are also equivalent. Most of his calculations rely on the vertical variation of the space variable, the values of X, M and Y. I found it easier to do the calculations by using the horizontal variation of the time variable t, using only X itself. Having recovered several of his key formulas this way, I looked for something new to do; so at the suggestion of my colleague D. Iglehart, computed the exact distribution of the maximum of the (positive) excursion straddling t, conditioned on the location and duration. This turns out to be expressible by a theta function and its derivative. I could not verify its monotonicity and asked Iglehart to plot it on a computer. Due to faulty transmission by telephone, a slight error in the plotted formula led to a curve decidedly not monotone. After the error was corrected the monotonicity was, of course, confirmed. Later I learned that the distribution had been found by N. H. Kuiper in a statistical test of random points on a circle. He became director of IHES, you know.

(II) The excursions of Brownian motion bear remarkable resemblance to those of a Markov chain with a single sticky recurrent boundary (see my Strasbourg Lectures). By identifying the explicit formulas for excursions with the general ones in the chain theory, hidden meanings of certain quantities are revealed. This is because in the final analytic expressions factorization and cancellation have taken place without our notice. Here is an example. One of the deeper formulas for Markov chains is the last exit decomposition:

$$p_{ij}(t) = \int_0^t p_{ii}(s)\, g_{ij}(t-s)ds.$$

The analogue for Brownian motion is (for $y > 0$)

$$p(t;\ 0,y) = \int_0^t p(s;\ 0,0)g(t-s,0,y)ds \tag{6}$$

which is identical to the first entrance formula owing to symmetry. We have

$$p(s;\ 0,0) = \frac{1}{\sqrt{2\pi s}}, \quad g(t;\ 0,y) = \frac{y}{\sqrt{2\pi t^3}} e^{-\frac{y^2}{2t}}.$$

Putting these quantities together we obtain for the right member of (6):

$$\begin{aligned} &\int_0^t \frac{1}{2\pi\sqrt{s(t-s)}} \frac{y}{t-s} e^{-y^2/(t-s)} ds \\ &= \int_0^t P(\gamma(t) \in ds) P\left(|X(t)| \in dy \,\middle|\, \gamma(t) = s\right) ds \end{aligned} \tag{7}$$

where $\gamma(t)$ is the last zero of X before t. Both probabilities in the last-written integral are derived by Lévy by his methods. The first is the arcsin law made famous by Feller's propaganda (mentioned by Lévy in his *Notice sur les travaux*). The second apparently was not understood by other authors until my 1976 paper which resulted from my attempt to unravel it. It is Theorem 42.5 in Lévy's book and it plays a key role in his *étude profonde.*[1]

(III) However pretty those excursions may be, Lévy's grand scheme is to string them all together on a new time scale, the local time at zero, and recover the Brownian motion as a Poisson point process run by the local clock. An illustration of this idea is his proof of the following theorem, the *pièce de resistance* of his conception of "*mesure du voisinage*", later known as local time.

Theorem. *We have almost surely*

$$\lim_{\epsilon \downarrow 0} \frac{1}{2\epsilon} m\left(\{s \leq t : \ |X(t)| < \epsilon\}\right) = L(t)$$

where m is the Lebesgue measure and $L(t)$ is the local time at t.

Recall that Lévy defined $L(.)$ by an inversion of a strictly increasing purely jumping stable process of index 1/2. It is a profound analogue of his construction of the singular Markov chain discussed earlier.

Lévy's proof, given at the end of his great paper *Sur certains processus stochatiques homogènes, Compostio math.* **7**, 1939[2], and not reproduced in his book (1948/1965), runs as follows. The total occupation time of $(0, \epsilon)$ by $|X|$ up to time t is the sum of the same occupation time $u_\epsilon(\varphi)$ during all the excursions φ up to local time $L(t)$:

$$\sum_{\varphi} u_\epsilon(\varphi) = \int_{[0,L(t)]\times R_+} u_\epsilon dN \tag{8}$$

Here N is the Poisson point process of the excursions, whose mean measure is the Lévy measure of the inverse local time, namely the stable process with exponent 1/2, given explicitly by Lévy in his work on "Lévy Processes". The expectation of the right member in (8) can be computed since we can compute the occupation time during an excursion, and there is independence between disjoint excursion intervals. (It has a neat density.) We need also an estimate for the second moment, which is supplied in my paper (dedicated to Lévy). Now under very general conditions on the first two moments of a Poisson sum like that in (8), the value of the random sum is asymptotically equivalent to its expectation in the limit, here as $\epsilon \downarrow 0$. Thus the theorem is proved by a straightforward computation of the expectation (a nice integral), and an adequate bound for the second moment, exactly as in the grand tradition of classical probability.

The same method gives quick proofs of a number of similar but easier results: the downcrossing result and Kingman's Cesàro mean result, etc. (Notes by A. A. Balkema on this topic exist.) Despite later alternative approaches to these matters, Lévy's original way should be preserved, not as a museum piece, but as a monument conjuring up the past and beckoning to the future.

Footnotes

(1) Lévy's grand tradition of deriving a wealth of explicit formulas has been continued in the recent work by Biane and Yor.

(2) It is regrettable that this paper was not cited in the volume by Dellacherie-Meyer mentioned above, but I was pleased to see it publicized in the exhibition at *École polytechnique* during the conference. When I first met Lévy in 1950, I asked him for a reprint of this paper and was told that all his papers were burned by the Nazis. His treatise *Theorie de l'addition des variables aléatoires* was not accessible to me during the war because library collections were stored in underground caves in China to escape from Japanese bombing.

Postscript.

This is the script of my talk delivered on June 22, 1987, at the Paul Lévy Colloque in École Polytechnique, Palaiseau. Some mathematical details were omitted in the oral version and a few impromptu remarks were added afterwards. A funny thing happened on its way to publication (which can only happen in Paris?). Although the talk was a pre-announced invited address, and I had told Laurent Schwartz from the outset that it would be an essay on Lévy's work, as the long title made abundantly clear, the solicited manuscript was "refereed". The so-called referee made two substantive points in his critique. First he proudly pointed out that the sample functions in Lévy's Columbus egg example are increasing on one hand, and have (numerous) discontinuities of the second kind on the other! Poor anonym, he must indeed be "tantalized" (see text) to have made this discovery, au niveau du deuxième cycle, and he would have worse trouble to read Lévy's original grand memoir (1951), which contains more bizarre things. But would he have the temerity to consult either my exposition of it (made easy for the humbler reader), or with Jacques Neveu or Paul André Meyer who could have disillusioned him? Mr. Meyer had heard me expound such matters in 1967 for several months, and apparently did eventually so instruct the anonym (rumors had it). The second complaint by the anonym was that in my discussion of Lévy's theorem on local time, no mention was made of a posterior work by "un japonais, il est vrai" (anonym's French). But my point in discussing that original proof (not reproduced in Lévy's 1948 book) was precisely that it had been buried for nearly forty years until it was resurrected (with a little dusting off) in my paper on Brownian excurision (1976), dedicated to his memory. That proof was not mentioned in the popular book by Ito and McKean, although five alternatives were putatively presented there. If one may judge from their own avowal, Lévy's original argument was not fully understood (this was not the first time it happened to Lévy), and only didactic prudence prevented me from saying so explicitly. The omission is now being redressed by the paper by Balkema and myself in this volume (or next). As I said toward the end of my talk, Lévy reduced his theorem to a matter of calculation with independent and identically distributed random variables, in the grand tradition of classical probability. Would the puzzled anonym have the curiosity now to take a look at that old stuff and savor a little of the time past? _Mais enfin_, this is what reminiscences (_souvenirs_) are all about.

There was another remark made by Meyer as well as the anonym, and that is I did not give exact references, so that the younger generation may not know what I am talking about. (I do wish that it would consult references, as the anonym did not.) As a matter of fact, when I sent the manuscript I wrote the secretaire that I would supply the references after it won official sanction. Now however, more than a year has passed and I must retract that promise on the excuse that the unexpected and unwanted polemique (inevitably politically motivated according to one participant, and only partially recounted above in order to spare another participant) is already too long and too boring to sustain any further interest in the matter.

FINIS. K.L.C.

CONDITIONAL BROWNIAN MOTION, WHITNEY SQUARES AND THE CONDITIONAL GAUGE THEOREM

BY

M. CRANSTON

Abstract. Let (X, P^x) be Brownian motion killed at $\tau_D = \inf\{t > 0 : X_t \notin D\}$, D a domain in $\mathbb{R}^2$ and (X, P_z^x) this motion conditioned on $X_{\tau_D} = z$. For Kato class potentials q we show $E_z^x[\exp\{-\int_0^{\tau_D} q(X_s)ds\}]$ is bounded from zero and infinity with little or no assumption on the smoothness of the boundary.

In this paper the conditional gauge theorem for Kato class potentials and most bounded planar domains is proved. This is done by revising a result of B. Davis on the occupation time of conditional Brownian motion in Whitney squares which will now be described. Denote by (X, P^x) Brownian motion killed on exiting a planar Greenian domain D and by (X, P_z^x) this motion conditioned to exit D at a point z. For the terminal points z one selects $z \in \Delta_1$ where Δ_1 is the minimal Martin boundary for D. The reader is referred to Doob (1983) for details on these so-called h-processes. Now consider a Whitney decomposition of D, $\{Q_j\}$. That is the Q_j are a family of squares with sides parallel to the coordinate axes, $Q_j^0 \cap Q_k^0 = \phi$ when $j \neq k$, $\text{diam}(Q_j) \leq \text{dist}(Q_j, D^c) \leq 4 \text{ diam } Q_j$ and $D = \bigcup_j Q_j$. If $\tau_D = \inf\{t > 0 : X_t \notin D\}$ set $T_Q = \int_0^{\tau_D} 1_Q(X_s)ds$. Then with $\sigma = \sigma_Q = \inf\{t > 0 : X_t \in Q\}$ and $|Q|$ denoting the area of Q, B. Davis (preprint) showed there is a universal positive constant c, so that if D is simply connected, $E_z^x T_Q \leq c|Q|P_z^x(\sigma < \tau_D)$. His argument can be quickly outlined: by

AMS 1980 Subject Classifications. Primary 60J45, 60J65.

the strong Markov property $E_z^z T_Q = E_z^z[E_z^{X_\sigma} T_Q; \sigma < \tau_D]$. If $z \in \Delta_1$ and letting $K(\cdot, z)$ be the minimal harmonic function with pole at z and $G(\cdot,\cdot)$ the Green function for $\frac{1}{2}\Delta$ on D then on $\{\sigma < \tau_D\}$

$$E_z^{X_\sigma} T_Q = \int_Q G(X_\sigma, y) K(y, z) K(X_\sigma, z)^{-1} dz \leq c \int_Q G(X_\sigma, y) dy$$

with the last inequality following from Harnack's inequality applied on Q to $K(\cdot, z)$. The last integral is $E^{X_\sigma} T_Q$. B. Davis now uses a clever argument involving the looping and scaling of Brownian motion, namely, $\sup_{w \in Q} P^w(T_Q > |Q|) \leq \epsilon$ where $1 - \epsilon$ is "the probability that standard Brownian motion makes a loop enclosing the disc of radius 5 diam (Q) about its starting point by time $|Q|$." Since D is simply connected and Q is a Whitney square such a loop must have exited D. Also the existence of such an ϵ is guaranteed by the scaling properties of planar Brownian motion. Thus $P^w(T_Q > m|Q|) \leq \epsilon^m |Q|$ so $E^w T_Q \leq c|Q|$ and Davis' result follows. For the purposes of proving the conditional gauge theorem an estimate on $E_z^z \int_0^{\tau_D} 1_Q(X_s)|q|(X_s)ds$ is needed where q is not necessarily in $L^\infty(D)$. This will be done by replacing the above argument involving loops by a Green function estimate. Namely, let Q be a Whitney square and $Q^* = \frac{3}{2}Q$. By aQ, $a > 0$, we mean the square concentric with Q dilated by the amount a. Then there is a positive constant c (independent of Q) such that

$$(1) \qquad G(w, y) \leq c, w \in \partial(Q^*), y \in Q\,.$$

This estimate holds for more than just simply connected domains. However, the inequality will not hold for all domains. Consider $D = \{z : 0 < |z| < 1\}$, then for Whitney squares Q near the origin the estimate $G(w, y) \leq c$ for $w \in \partial(Q^*)$, $y \in Q$ will not hold, $\{0\}$ is not a large enough set (it is polar) to hold down the Green function. The actual condition that will be imposed involves the logarithmic capacity of D^c near Q. For capacitable sets K, $C_\ell(K)$ denotes the logarithmic capacity of K (see Landkof (1972) for a good exposition). If K is a line segment of length a, $C_\ell(K) = \frac{a}{4}$. If $K = \{z : |z| < a\}$ then $C_\ell(K) = a$.

Using $f \simeq g$ to mean a two-sided inequality $c_1 f(x) \leq g(x) \leq c_2 f(x)$ holds with independent positive constants c_1 and c_2 the condition that is imposed is the following:

$$(2) \qquad C_\ell((2+8\sqrt{2})Q \setminus \frac{3\sqrt{2}}{2} Q \cap D^c) \simeq \operatorname{diam} Q\,, Q \in W\,.$$

A bit of arithmetic shows $\partial D \cap (\frac{3\sqrt{2}}{2} Q)^0 = \phi$ but $\partial D \cap (1+8\sqrt{2})Q \neq \phi$. Thus the square Q is expanded a little more to $(2+8\sqrt{2})Q$ to capture a significant piece of D^c.

For D simply connected, Lemma 1 below implies condition (2) holds. For domains such as Salisbury's maze (Salisbury (1986)) property (2) holds. The latter domain which looks like this

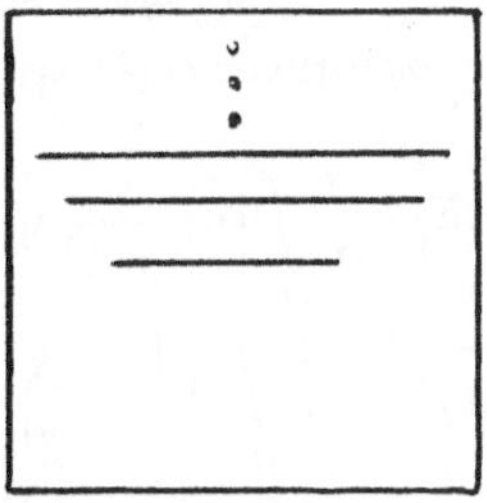

is not simply connected. It seems that the conditional gauge theorem should hold for all bounded planar domains but at this point I don't know how to get around using condition (2).

We now study the Green function for an unbounded planar domain. More specifically, let E be a compact set with logarithmic capacity $C_\ell(E) > 0$. Then the Green function for $F = \mathbf{R}^2 \setminus E$ exists and will be denoted by $G_F(x,y)$. We now give an expression for G_F by paraphrasing Landkof (1972). First, $\lim_{y\to\infty} G_F(x,y) = G_F(x,\infty)$ exists and if w_F^x is the harmonic measure for F then

$$G_F(x,y) = \frac{1}{\pi} \ell n \frac{1}{|x-y|} - \frac{1}{\pi} \int_{\partial F} \ell n \frac{1}{|z-y|} w_F^x(dz) + G_F(x,\infty)\,.$$

Thus, fixing a point $x_0 \in E$,

$$G_F(x,\infty) = G_F(x,y) - \frac{1}{\pi}\ell n\frac{|x-x_0|}{|x-y|} + \frac{1}{\pi}\int_{\partial F} \ell n\frac{1}{|z-y|}w_F^x(dz) + \frac{1}{\pi}\ell n|x-x_0|.$$

Since the first three terms on the right hand side are bounded as $x \to \infty$ it follows that $G_F(x,\infty) = \frac{1}{\pi}\ell n|x-x_0| + O(1)$ as $x \to \infty$. Since there is only one boundary point at ∞ and $G_F(x,\infty) - \frac{1}{\pi}\ell n|x-x_0|$ is bounded and harmonic off E, $\lim_{x\to\infty}[G_F(x,\infty) - \frac{1}{\pi}\ell n|x-x_0|] = L$ exists. Thus

$$\lim_{x\to\infty}\frac{1}{\pi}\int_{\partial F} \ell n\frac{1}{|z-y|}w_F^x(dz) = L - G_F(\infty,y).$$

Also, $w_F^x(dz) \to \lambda(dz)$ as $x \to \infty$ so $U^\lambda(y) = \frac{1}{\pi}\int_{\partial F}\ell n\frac{1}{|z-y|}\lambda(dz) = L - G_F(\infty,y)$. Now $U^\lambda(y) = L$ for $y \in E$ and $U^\lambda(y) \le L$ otherwise. This identifies L as $-\ell n\ C_\ell(E)$ where $C_\ell(E)$ is the logarithmic capacity of E. Consequently,

$$\begin{aligned} G_F(y,\infty) &= -\ell n\ C_\ell(E) - \frac{1}{\pi}\int_{\partial F}\ell n\frac{1}{|z-y|}\lambda(dz) \\ &= \frac{1}{\pi}\ell n\frac{|x-x_0|}{C_\ell(E)} - \frac{1}{\pi}\int_{\partial F}\ell n\frac{|x-x_0|}{|z-y|}\lambda(dz). \end{aligned}$$

and therefore,

$$(3)\quad G_F(x,y) = \frac{1}{\pi}\int_{\partial F}\ell n\frac{|z-y|}{|x-y|}w_F^x(dz) + \frac{1}{\pi}\ell n\frac{|x-x_0|}{C_\ell(E)} - \frac{1}{\pi}\int_{\partial F}\ell n\frac{|x-x_0|}{|z-x|}\lambda(dz).$$

Eventually, a bound on $G_F(x,y)$ for certain values of x and y will be needed so $C_\ell(E)$ will need to be estimated. According to Landkof (1972) this may be calculated by means of transfinite diameter. Set

$$\ell n\frac{1}{d_n(E)} = \min_{x_1,\dots,x_n\in E}\frac{1}{\pi}\binom{n}{2}^{-1}\sum_{i<j}\ell n\frac{1}{|x_i-x_j|},$$

then $\lim_{n\to\infty} d_n(E) = d(E)$ is called the transfinite diameter of E and $d(E) = C_\ell(E)$. This lemma will be useful.

LEMMA 1. *If π is the projection onto a coordinate axis and E is compact then $C_\ell(\pi E) \leq C_\ell(E)$.*

PROOF: Let $\xi_1, \ldots, \xi_n \in E$ be such that $\ell n \frac{1}{d_n(\pi E)} = \frac{1}{\pi}\binom{n}{2}^{-1} \sum_{i<j} \ell n \frac{1}{|\pi\xi_i - \pi\xi_j|}$. Since $|\pi\xi_i - \pi\xi_j| \leq |\xi_j - \xi_j|$ it follows that

$$\ell n \frac{1}{d_n(\pi E)} \leq \frac{1}{\pi}\binom{n}{2}^{-1} \sum_{i<j} \ell n \frac{1}{|\xi_i - \xi_j|} \leq \ell n \frac{1}{d_n(E)} .$$

This implies the lemma. ∎

For Q_j a Whitney square, $2Q_j \subset D$, so Harnack's inequality applies on $Q_j^* = \frac{3}{2}Q_j$ to positive harmonic functions on D. It is important to observe that $Q_j^* \subset 2Q_j \subset D$ implies the constant in Harnack's' inequality is independent of Q_j.

THEOREM 2. *Suppose D is a bounded planar domain and $W = \{Q_j\}_{j\geq 0}$ is a Whitney decomposition for D. If $C_\ell((2+8\sqrt{2})Q \setminus \frac{3\sqrt{2}}{2}Q) \cap D^c) \simeq$ diam Q for every $Q \in W$ then there is a positive constant c independent of diam Q such that for $x \in \partial Q^* = \partial(\frac{3}{2}Q)$ and $y \in Q$, $G(x,y) \leq c$.*

PROOF: For $Q \in W$, set $E = E(Q) = ((2+8\sqrt{2})Q \setminus \frac{3\sqrt{2}}{2}Q) \cap D^c$ and $F = \mathbf{R}^2 \setminus E$. Then E being nonpolar implies G_F exists and $F \supset D$ so $G_F(x,y) \geq G(x,y)$ for $x, y \in D$. By (3)

$$G_F(x,y) = \frac{1}{\pi}\int_{\partial F} \ell n \frac{|z-y|}{|x-y|} w_F^x(dz) + \frac{1}{\pi}\ell n \frac{|x-x_0|}{C_\ell(E)} + \frac{1}{\pi}\int_{\partial F} \ell n \frac{|z-x|}{|x-x_0|}\lambda(dz),$$

where $x_0 \in E$ is fixed. However for $x_0, z \in E$, $x \in \partial Q^*, y \in Q$ there exist positive constants c_0 and c_1 independent of diam Q such that $|z-y| \leq c_1$ diam Q, $|x-y| \leq c_0$ diam Q, $|x-z| \leq c_1$ diam Q, $|x-x_0| \simeq$ diam Q, $C_\ell(E) \geq c_0$ diam Q, $w_F^x(E) = 1$, $\lambda(E) = 1$. Thus $G_F(x,y) \leq c$ and so $G(x,y) \leq c$ for some positive constant c independent of diam Q. ∎

COROLLARY 3. *Suppose D is a simply connected planar Greenian domain. Let $W = \{Q_j\}_{j\geq 0}$ be a Whitney decomposition for D. There is a positive constant c such that for every $Q \in W$, $x \in \partial Q^*$, $y \in Q$, $G(x,y) \leq c$.*

PROOF: Set $E = E(Q) = ((2+8\sqrt{2})Q \setminus \frac{3\sqrt{2}}{2}Q) \cap D^c$. Then the projection πE onto one or the other axis must contain a line segment of length proportional to diam Q. Now use Lemma 2. ∎

COROLLARY 4. *Suppose D is a bounded planar domain which satisfies (2). There is a positive constant c such that if $q \in L^1(D)$ then*

$$E_z^x\left[\int_0^{\tau_D} 1_Q|q|(X_s)ds\right] \leq c\int_Q |q|(y)dy P_z^x(\sigma^* < \tau_D) \tag{4}$$

for $Q \in W$, $x \in D \setminus Q^*$, $z \in \Delta_1$, $\sigma^* = \inf\{t > 0 : X_t \in Q^*\}$.

PROOF: For $Q \in W$, $x \in D \setminus Q^*$, $z \in \Delta_1$,

$$\begin{aligned} E_z^x\left[\int_0^{\tau_D} 1_Q \cdot |q|(X_s)ds\right] &= E_z^x\left[E_z^{X_{\sigma^*}}\left[\int_0^{\tau_D} 1_Q \cdot |q|(X_s)ds\right];\sigma^* < \tau_D\right] \\ &= E_z^x\left[\int_Q G(X_{\sigma^*},y)K(y,z)K(X_{\sigma^*},z)^{-1}|q|(y)dy;\ \sigma^* < \tau_D\right] \\ &\leq c\int_Q |q|(y)dy\ P_z^x(\sigma^* < \tau_D), \end{aligned}$$

by Harnack and Theorem 2. ∎

The following is less general than in Cranston, McConnell (1983) where the estimate $E_z^x\tau_D \leq c|D|$ was proved by other means.

COROLLARY 5. *There is a positive constant c such that if D is a planar domain satisfying (2) then $E_z^x\tau_D \leq c|D|$ for $x \in D$, $z \in \Delta_1$.*

PROOF: Suppose $x \in Q_0$ as we may and that $x \in Q_j^*$ for $j = 1, \ldots, N$ (N is finite and is independent of x, D and Q_0.) Then with $\tilde{Q} = \bigcup_{j=0}^N Q_j$

$$E_z^x \tau_D = E_z^x T_{\tilde{Q}} + \sum_{j=N+1}^{\infty} E_z^x T_{Q_j}$$

$$\leq E_z^x T_{\tilde{Q}} + c \sum_{j=N+1}^{\infty} |Q_j| P_z^x(\sigma_j^* < \tau_D)$$

where $\sigma_j^* = \inf\{t > 0 : X_t \in Q_j^*\}$. Letting

$$\tilde{Q}^* = \bigcup_{j=0}^{N} Q_j^* \,, \tau^* = \inf\{t > 0 : X_t \notin \tilde{Q}^*\}$$

$$E_z^x T_{\tilde{Q}} \leq E_z^x \tau^* + \sum_{j=0}^{N} E_z^x [E_z^{X_{\tau^*}} T_{Q_j}]$$

$$= \int_{\tilde{Q}^*} G_{\tilde{Q}^*}(x,y) K(y,z) K(x,z)^{-1} dy + \sum_{j=0}^{N} E_z^x \left[E_z^{X_{\tau^*}} \left[E_z^{X_{\sigma^*}} T_{Q_j}; \sigma_j^* < \tau_D\right]\right]$$

$$\leq c E^x \tau_{\tilde{Q}^*} + c \sum_{j=0}^{N} |Q_j| \,,$$

by Harnack and (4) with $q \equiv 1$. Now it is known (see Cranston, McConnell (1983)) that $E^x \tau_{\tilde{Q}^*} \leq c|\tilde{Q}^*|$. The result now follows since $|D| = \sum_{j=0}^{\infty} |Q_j|$. ∎

The conditional gauge arises when applying the Feynman-Kac formula to solve $(-\frac{1}{2}\Delta + q)u = 0$ on $D, u = f$ on ∂D. Namely, if $e_q(t) = \exp\{-\int_0^t q(X_s) ds\}$, $t \geq 0$, then $u(x)$, if it exists, (see Chung, Li, Williams (1983)) is given by $u(x) = E^x[f(X_{\tau_D -}) e_q(\tau_D)] = \int_{\Delta_1} f(z) E_z^x[e_q(\tau_D)] P^x(X_{\tau_D -} \in dz)$. The quantity $u(x,z) = E_z^x[e_q(\tau_D)]$ is the so-called conditional gauge. The most natural condition on q in two dimensions is the Kato condition

$$\lim_{r \to 0} \sup_{x \in D} \int_{|x-y|<r} |q|(y) \ell n \frac{1}{|x-y|} dy = 0$$

written $q \in K_2(D)$ (in dimensions $d \geq 3$, $\ell n \frac{1}{|x-y|}$ is replaced by $|x-y|^{2-d}$.) Falkner (1983) first proved a conditional gauge theorem, namely, that $u(x,z)$ is bounded from zero and infinity when it is not identically infinite, for $q \in L^\infty(D)$ and D a C^2 domain. Later Zhao (1983, 1984) relaxed the condition on q to Kato rather than bounded. Lately, Cranston, Fabes, Zhao (1988) showed the conditonal gauge theorem holds for Kato q, bounded Lipschitz domains in $d \geq 3$ and $\frac{1}{2}\Delta$ may be replaced by a uniformly elliptic divergence form operator. Falkner (1987) pointed out that an example in Cranston, McConnell (1983) implies the conditional gauge theorem even for bounded q can not hold for bounded domains when $d \geq 3$ without some smoothness assumptions about the boundary. In fact, in the example cited a bounded three dimensional domain D and minimal Martin boundary point z were given such that $P_z^x(\tau_D = \infty) = 1$. However, in two dimensions $E_z^x \tau_D \leq c|D|$. Below we show the conditional gauge theorem holds for bounded planar domains which satisfy (2). This was established by Zhao (1987) for bounded two dimensional Jordan domains.

McConnell (1988) has extended this to bounded simply connected domains. Both these results are proved by getting the so-called $3G$ inequality. It seems likely this result should hold for bounded planar domains. An example in the thesis of Northshield shows $|D| < \infty$ will not suffice without the additional assumption $q \in L^1(D)$ as well as $q \in K_2(D)$. These conditional gauge results are thus tightly linked (and maybe even equivalent?) with uniform bounds $E_z^x \tau_D \leq c$.

THEOREM 6. *Suppose D is a bounded planar domain satisfying (2) and $q \in K_2(D)$. Then if $E_{z_0}^{x_0}[e_q(\tau_D)] < \infty$ for some $x_0 \in D$, $z_0 \in \Delta_1$ there exist positive constants c_1 and c_2 such that $c_1 \leq E_z^x[e_q(\tau_D)] \leq c_2$ for all $x \in D$, $z \in \Delta_1$.*

PROOF: In Chung (1985) it was pointed out that only a basic estimate need be established. To that end as in Chung (1985) introduce subdomains $D_1, D_2 \subset D$ with $\overline{D}_1 \subset D_2$, $\overline{D}_2 \subset D$, $C = D \setminus D_1$ is connected and $\partial D \subset \partial C$. Then all that needs be shown is that given $\epsilon < 1$ there is a δ such that if $|C| < \delta$

then $E_z^x\left[\int_0^{\tau_C}|q|(X_s)ds\right] \le \epsilon$ for $x \in \partial D_2$, $z \in \Delta_1$. Let $W = \{Q_j\}$ be a Whitney decomposition for D. Notice $G(x,y) = \ell n\frac{1}{|x-y|} - \int_{\Delta_1}\ell n\frac{1}{|z-y|}w_D^x(dz) = \ell n\frac{\text{diam } D}{|x-y|} - \int_{\Delta_1}\ell n\frac{\text{diam } D}{|z-y|}w_D^x(dz)$ and diam $D \ge |z-y|$ for $z, y \in \overline{D}$ so $G(x,y) \le \ell n\frac{1}{|x-y|} + |\ell n \text{ diam } D| = \ell n\frac{1}{|x-y|} + c_1$. Also since $q \in K_2(D)$ and D is bounded it follows that $q \in L^1(D)$. Set $\tilde{Q} = \cup\{Q_j : x \in Q_j^*\}$, $\tilde{Q}^* = \cup\{Q_j^* : x \in Q_j^*\}$, $\tau^* = \inf\{t > 0 : X_t \notin \tilde{Q}^*\}$. Let θ_t denote the shift operator on paths, $X_s \circ \theta_t = X_{t+s}$.

Then for $x \in \partial D_2, z \in \Delta_1$,

$$
\begin{aligned}
E_z^x\left[\int_0^{\tau_C}|q|(X_s)ds\right] &= \sum_{j=0}^{\infty}E_z^x\left[\int_0^{\tau_C}|q|1_{Q_j\cap C}(X_s)ds\right] \\
&\le E_z^x\left[\int_0^{\tau^*}|q|1_{\tilde{Q}^*\cap C}(X_s)ds\right] + \sum_{j=0}^{N}E_z^x\left[E_z^{X_{\tau^*}}\left[\int_0^{\tau_C}|q|1_{Q_j\cap C}(X_s)ds\right]\right] \\
&\quad + \sum_{j=N+1}^{\infty}E_z^x\left[E_z^{X_{\sigma_j^*}}\left[\int_0^{\tau_C}|q|1_{Q_j\cap C}(X_s)ds\right];\sigma_j^* < \tau_D\right] \\
&\le \int_{\tilde{Q}^*\cap C}G(x,y)K(y,z)K(x,z)^{-1}|q|(y)dy \\
&\quad + \sum_{j=0}^{N}E_z^x\left[E_z^{X_{\tau^*}}\left[E_z^{X_{\sigma_j^*}}\left[\int_0^{\tau_C}|q|1_{Q_j\cap C}(X_s)ds\right];\theta_{\tau^*}\circ\sigma_j^* < \tau_D\right]\right] \\
&\quad + c\sum_{j=N+1}^{\infty}\int_{Q_j\cap C}|q|(y)dyP_z^x(\sigma_j^* < \tau_D) \text{ by (4)} \\
&\le c\int_{\tilde{Q}^*\cap C}G(x,y)|q|(y)dy + c\sum_{j=0}^{\infty}\int_{Q_j\cap C}|q|(y)dy\,, \text{ by Harnack and (4)} \\
&\le c\int_{\tilde{Q}^*\cap C}\ell n\frac{\text{diam } D}{|x-y|}|q|(y)dy + c\int_C|q|(y)dy\,.
\end{aligned}
$$

Since $q \in K_2(D)$ implies $q \in L^1(D)$, this upper bound can be made arbitrarily small by making $|C|$ small. ∎

Remarks.

1. With a slightly improved argument one can show that if D satisfies (2), $|D| < \infty$ and $q \in K_2(D) \cap L^1(D)$ that Theorem 6 will hold.

2. The above results all hold if $\frac{1}{2}\Delta$ is replaced by a divergence form uniformly elliptic operator $A = \frac{\partial}{\partial x_i}(a_{ij}(x)\frac{\partial}{\partial x_j})$ with bounded measurable coefficients.

This follows since the Green function for A on D is equivalent to the Green function for $\frac{1}{2}\Delta$ on D by a result of Littman, Stampacchia, Weinberger (1963).

3. For another use and perhaps the first of Whitney squares and conditional Brownian motion, see Bañuelos (1987).

References

R. Bañuelos (1987). On an estimate of Cranston and McConnell. *Prob. Th. Rel. Fields,* **76**, 311–323.

K.L. Chung (1985). The gauge and conditional gauge theorem. *Sém. de Prob.* **XIX**, 1983/84. *Lecture Notes in Math.* **1123**, 496–503.

K.L. Chung, P.Li, R. Williams (1986). Comparison of probability and classical methods for the Schrödinger equation. *Exp. Math. Band 4, Heft 3,* 271–278.

M. Cranston, E. Fabes, Z. Zhao (1988). Potential theory for the Schrödinger equation. To appear, Trans. Amer. Math. Soc.

M. Cranston, T.R.McConnell (1983). The lifetime of conditioned Brownian motion. *Z. Wahr. ver.Geb.* **70**, 1–11.

B. Davis (1987). Conditioned Brownian motion in planar domains. Preprint.

J.L. Doob (1983). *Classical Potential Theory and Its Probabilistic Counterpart.* Springer-Verlag, Berlin.

N. Falkner (1983). Feynman-Kac functionals and positive solutions of $\frac{1}{2}\Delta u + qu = 0$. *Z.Wahr. ver. Geb.* **65**, 19–33.

N. Falkner (1987). Conditional Brownian motion in rapidly exhaustible domains. *Annals of Prob., Vol. 15,* **4**, 1501–1514.

N.S. Landkof (1972). *Foundations of Modern Potential Theory.* Springer-Verlag, Berlin.

W.Littman, G. Stampacchia, H. Weinberger (1963). Regular points for elliptic equations with discontinuous coefficients. *Ann.Scuola Norm. Sup. Pisa, Serie III, XVII Fasc. I–II*, 45–79.

T. R. McConnell (1988). Oral communication.

T. Salisbury (1986). A Martin boundary in the plane, *Trans. Amer. Math.Soc., Vol. 293, No.2*, 623–642.

Z. Zhao (1983). Conditional gauge with unbounded potential. *Z. Wahr. ver. Geb.* **63**, 13–18.

Z. Zhao (1984). Uniform boundedness of conditional gauge and Schrödinger equations. *Comm. Math. Physics* **93**, 19–31.

Z. Zhao (1987) Green functions and conditioned gauge theorem for a two-dimensional domain, *Seminar on Stochastic Processes*, Birkhäuser, Boston, 283–294.

Michael Cranston
University of Rochester
Mathematics Department
Rochester, New York, 14627

LOCAL FIELD GAUSSIAN MEASURES

by

STEVEN N. EVANS

1. Introduction

A pervasive undercurrent in the study of Gaussian measures is that they are the class of probability measures which it is natural to study if one requires that we see probabilistic properties which are consonant with the linearity and orthogonality properties of the spaces on which the measures are defined. For instance, one entry point into the theory of Gaussian random variables on an arbitrary real vector space with suitable measurable structure is to define a random variable X as being Gaussian if whenever X_1, X_2 are two independent copies of X, then the pair $(\alpha_{11}X_1+\alpha_{12}X_2, \alpha_{21}X_1+\alpha_{22}X_2)$ has the same law as (X_1, X_2) for each pair of orthonormal vectors $(\alpha_{11}, \alpha_{12}), (\alpha_{21}, \alpha_{22}) \in \mathbb{R}^2$. It can be shown that, in the appropriate special cases, this abstract definition is equivalent to the usual concrete definitions for $\mathbb{R}^n$-valued Gaussian random variables and Gaussian stochastic processes.

In this paper we consider vector spaces over fields other than $\mathbb{R}$ or $\mathbb{C}$, namely the so-called local fields (a topological field is said to be a local field if it is locally compact, non-discrete, and totally disconnected, see §2). We propose an answer to the question, "What is the appropriate analogue for the class of Gaussian measures on these objects?" There is a suitable concept of orthogonality in the local field setting (see §3), and so our approach is to begin with the local field analogue of the abstract definition given above, and then see where that leads us. We can establish a body of results which in many particulars resembles the usual theory. For example, linear transformations of "Gaussian" variables are "Gaussian" and orthogonality is equivalent to independence (see §7). We also observe the sort of zero-one behavior that we have come to expect from the Gaussian theory (see §6).

There are even results which have no Gaussian antecedents. For instance, on spaces with enough structure, it turns out that the "Gaussian" random variables are the class that is imposed upon us if we simply insist that linear combinations of independent random variables drawn from the class behave appropriately (see Corollary 7.4). This characterization is, a priori, much weaker than our original definition in terms of the effect of orthogonal transformations and has no counterpart in the Euclidean theory.

Unfortunately, we also lose something in the new setting. Roughly speaking, the local field notion of orthogonality is an L^{∞} object rather than an L^{2} object. As a consequence, the resulting "Gaussian" theory is not a second-order theory where some analogue for the concept of covariance describes the distributional picture. In particular, the law of a "Gaussian" process is not described by its family of two-dimensional marginal distributions (see Example 8.1). Also, while we show that there are broad classes of random series of functions which are stationary, we also find that there is no counterpart to the representation of a general stationary Gaussian process on the circle as a random Fourier series (see §9).

2. Local Fields

This section is essentially a summary of selected results from [Taibleson, 1975] and [Schikhof, 1984]. We refer the reader to these works for a fuller account.

Let K be a topological field. That is, K is a field such that the additive and multiplicative groups of K are both topological groups. Suppose that K is locally compact and non-discrete. If K is connected, then K is either $\mathbb{R}$ or $\mathbb{C}$. If K is disconnected, then K is totally disconnected and we say that K is a local field.

From now on, we let K be a fixed local field. There is a real-valued mapping on K which we denote by $x \mapsto |x|$. The set of values taken by this map is the set $\{q^k: k \in \mathbb{Z}\} \cup \{0\}$, where $q = p^c$ for some prime p and positive integer c, and has the following properties:

$$|x| = 0 \iff x = 0; \qquad (2.1)$$

$$|xy| = |x|\,|y|; \qquad (2.2)$$

$$|x+y| \le |x| \vee |y|. \qquad (2.3)$$

A map with properties (2.1)-(2.3) is called a non-archimedean valuation. Property (2.3) is known as the ultrametric inequality. The mapping $(x,y) \mapsto |x-y|$ on $K \times K$ is a metric on K which gives the topology of K. A consequence of the ultrametric inequality is that if $|x| \neq |y|$, then $|x+y| = |x| \vee |y|$. This latter result implies that for every "triangle" $\{x,y,z\} \subset K$ we have that at least two of the lengths $|x-y|$, $|x-z|$, $|y-z|$ must be equal and is therefore often called the isosceles triangle property.

In the words of [Schikhof, 1984], "... we shall follow a bad but widespread habit and omit the subscript K in 1_K and n_K (:= the sum of n times 1_K)." Clearly, $|1| = 1$. If we choose $\rho \in K$ so that $|\rho| = q^{-1}$, then

$$\rho^k\{x: |x| \le 1\} = \{x: |x| \le q^{-k}\}$$
$$= \{x: |x| < q^{-(k-1)}\}$$

for each $k \in \mathbb{Z}$. Each of the sets $\{x: |x| \le q^k\}$, $k \in \mathbb{Z}$, is an additive subgroup of K and for $\ell < k$ the quotient group $\{x: |x| \le q^k\}/\{x: |x| \le q^\ell\}$ has order $q^{k-\ell}$. If $\{a_1,\ldots,a_q\}$ is a complete list of coset representatives of $\{x: |x| \le q^{-1}\}$ in $\{x: |x| \le 1\}$, then we may express each $y \in \{x: |x| \le 1\}$ uniquely as $y = \sum_{k=0}^{\infty} b_k \rho^k$, where $b_k \in \{a_1,\ldots,a_q\}$ for each k.

EXAMPLE 2.1. Fix a positive prime p. We can write $r \in \mathbb{Q}\setminus\{0\}$ uniquely as $r = p^s(a/b)$ where a and b are not divisible by p, and we set $|r| = p^{-s}$. If we set $|0| = 0$, then the map $|\cdot|$ has the properties (2.1)-(2.3). The map

$(x,y) \mapsto |x-y|$ defines a metric on $\mathbb{Q}$ and we denote the completion of $\mathbb{Q}$ in this metric by $\mathbb{Q}_p$. The field operations on $\mathbb{Q}$ and the map $|\cdot|$ may be uniquely extended to $\mathbb{Q}_p$ to make $\mathbb{Q}_p$ a local field, the so-called field of p-adic numbers. In the above notation we have $q = p$, and we may take $\rho = p$. The subgroup $\mathbb{Z}_p = \{x \in \mathbb{Q}_p: |x| \leq 1\}$ is both the closure of $\mathbb{Z}$ and $\mathbb{N}$ in $\mathbb{Q}_p$. For each $n \in \mathbb{N}$, the set $\{0,1,\ldots,p^n-1\}$ is a complete list of coset representatives of $p^n\mathbb{Z}_p$ in $\mathbb{Z}_p$. In particular, we can write $x \in \mathbb{Z}_p$ as

$$x = \Sigma_{k=0}^{\infty} b_k p^n,$$

where $b_k \in \{0,1,\ldots,p-1\}$ for all k.

Returning to the general theory, there is a unique Borel measure μ on K for which

$$\mu(x+A) = \mu(A), \qquad x \in K,$$

$$\mu(xA) = |x|\mu(A), \quad x \in K,$$

and

$$\mu(\{x: |x| \leq 1\}) = 1,$$

(the measure μ is a suitably normalised Haar measure on the additive group of K).

There is a character χ on the additive group of K with the properties

$$\chi(\{x: |x| \leq 1\}) = \{1\},$$

and

$$\chi(\{x: |x| \leq q\}) \neq \{1\}.$$

For $N = 1,2,\ldots$, the correspondence $\lambda \leftrightarrow \chi_\lambda$, where $\chi_\lambda(x) = \chi(\lambda\cdot x)$, establishes an isomorphism between the additive group of K^N and its dual. The uniqueness theorem for Fourier transforms in this setting reads as follows.

LEMMA 2.3. If ν_1, ν_2 are two finite measures on K^N such that

$$\int \chi(\lambda \cdot x)\nu_1(dx) = \int \chi(\lambda \cdot x)\nu_2(dx)$$

for all $\lambda \in K^N$, then $\nu_1 = \nu_2$.

There is one Fourier transform which is of particular interest to us. If Φ: $[0,\infty[\to \{0,1\}$ is the indicator function of the interval $[0,1]$, then

$$q^{-n}\int_K \chi(\lambda x)\Phi(q^{-n}|x|)\mu(dx) = \Phi(q^n|\lambda|). \qquad (2.4)$$

We remark that Φ has the property

$$\Phi(a \vee b) = \Phi(a)\Phi(b), \quad a,b \geq 0. \qquad (2.5)$$

The valuation on K can be uniquely extended to a non-archimedean valuation on the algebraic closure of K. The completion of the algebraic closure in the metric defined by the valuation is also algebraically closed. We let $\mathbb{C}_p$ denote the completion of the algebraic closure of $\mathbb{Q}_p$. The field $\mathbb{C}_p$ is not a local field.

For want of somewhere better to include them, we finish this section with three technical lemmas that will be needed in §9.

LEMMA 2.6. Suppose that $a \geq b \geq 0$. If $\alpha, \beta \in K$, then

$$(|\alpha|a) \vee (|\alpha+\beta|b) = (|\alpha|a) \vee (|\beta|b).$$

PROOF. Suppose, first of all, that $|\alpha|a \geq |\beta|b$. From the ultrametric inequality, we have that

$$|\alpha+\beta|b \le (|\alpha| \vee |\beta|)b \le |\alpha|a,$$

and the equality holds.

On the other hand, if $|\alpha|a < |\beta|b$, then $|\alpha| < |\beta|$. Applying the isosceles triangle property, we have that $|\alpha+\beta| = |\beta|$, and the equality also holds. ■

LEMMA 2.7. Suppose that $a \ge b \ge c \ge 0$. If $\alpha_1,\ldots,\alpha_n \in K$, then

$$|\alpha_1|a \vee (|\alpha_1-(\alpha_2+\cdots+\alpha_n)| \vee |\alpha_2| \vee\cdots\vee |\alpha_{n-1}|)b \vee |\alpha_n|c$$
$$= |\alpha_1|a \vee (|\alpha_2| \vee\cdots\vee |\alpha_n|)b.$$

PROOF. The left-hand side is at most the right-hand side by the ultrametric inequality. The reverse inequality is clear except in the following cases. For ease of notation, we set $\beta = \alpha_1-(\alpha_2+\cdots+\alpha_n)$.

Case I.

$$|\alpha_1|a \vee (|\alpha_2| \vee\cdots\vee |\alpha_{n-1}|)b < |\beta|b \qquad (2.7.1)$$

$$|\alpha_n|c \le |\beta|b \qquad (2.7.2)$$

Case II.

$$|\alpha_1|a \vee (|\beta| \vee |\alpha_2| \vee\cdots\vee |\alpha_{n-1}|)b < |\alpha_n|c. \qquad (2.7.3)$$

Suppose that Case I holds. From (2.7.1) and the ultrametric inequality, we have that

$$|\alpha_1-(\alpha_2+\cdots+\alpha_{n-1})| \le (|\alpha_1| \vee\cdots\vee |\alpha_{n-1}|)$$
$$< |\beta|$$
$$= |(\alpha_1-(\alpha_2+\cdots+\alpha_{n-1}))-\alpha_n|.$$

The isosceles triangle property then implies that

$$|\alpha_n| < |\beta|,$$

so the reverse inequality holds.

Since

$$|\alpha_n| \le |\alpha_1| \vee \cdots \vee |\alpha_{n-1}| \vee |\beta|$$

by the ultrametric inequality, we see that Case II cannot hold. ∎

LEMMA 2.8. Suppose that $a \geq b \geq 0$. If $\alpha_1, \ldots, \alpha_n \in K$, then

$$(|\alpha_1| \vee \cdots \vee |\alpha_{n-1}|)a \vee |\alpha_n|b \vee |\alpha_1 + \cdots + \alpha_n|a$$
$$= (|\alpha_1| \vee \cdots \vee |\alpha_n|)a.$$

PROOF. The left-hand side is at most the right-hand side by the ultrametric inequality, and the reverse inequality is clear except when $|\alpha_n| > (|\alpha_1| \vee \cdots \vee |\alpha_{n-1}|)$. In this case, we have from the ultrametric inequality that $|\alpha_n| > |\alpha_1 + \cdots + \alpha_{n-1}|$, and so, by the isosceles triangle property,

$$|\alpha_1 + \cdots + \alpha_n| = |\alpha_n|,$$

and the result follows. ∎

3. Normed Spaces and Orthogonality

The material in this section is included for ease of reference and is a summary of results and ideas which may be found in [Schikhof, 1984].

DEFINITION 3.1. Let E be a vector space over K. A norm on E is a map $\| \ \|: E \to [0,\infty[$ such that

$$\|x\| = 0 \Leftrightarrow x = 0, \tag{3.2}$$

$$\|\lambda x\| = |\lambda| \, \|x\|, \ \lambda \in K, \tag{3.3}$$

$$\|x+y\| \leq \|x\| \vee \|y\|. \tag{3.4}$$

We call the pair $(E, \| \ \|)$ a normed vector space (over K). If E is complete in the metric $(x,y) \mapsto \|x-y\|$, we say that E is a Banach space (over K).

Property (3.4) is also called the ultrametric inequality and leads to the obvious analogue of the isosceles triangle property.

EXAMPLE 3.5. All normed finite-dimensional vector spaces over K are Banach spaces. In particular, for N = 1,2,..., the space $(K^N, | \ |)$, where

$$|(x_1,\dots,x_N)| = |x_1| \vee \cdots \vee |x_N|,$$

is a Banach space.

EXAMPLE 3.6. Let X be a compact topological space. The space $C(X \to K)$ of K-valued continuous functions on X equipped with the norm $\| \ \|_c$ given by

$$\|f\|_c = \sup\{|f(x)|: x \in K\}$$

is a Banach space.

EXAMPLE 3.7. Let $(\Omega, \mathcal{F}, P)$ be a probability space. Let L^∞ be the set of measurable functions $f: \Omega \to K$ such that $\text{ess sup}\{|f(\omega)|: \omega \in \Omega\} < \infty$. If we say that f = g whenever

$f(\omega) = g(\omega)$ for almost all ω, then L^∞ equipped with the norm $\| \ \|_\infty$ defined by

$$\|f\|_\infty = \text{ess sup}\{|f(\omega)|: \omega \in \Omega\}$$

is a Banach space. (This example is not in [Schikhof, 1984], but the proof proceeds just as in the parallel real case.)

DEFINITION 3.8. Suppose that $(E, \| \ \|)$ is a normed space (over K). We say that a set $D \subset E$ is orthogonal if for every finite subset $\{x_1, \ldots, x_n\} \subset D$ and each $\lambda_1, \ldots, \lambda_n \in K$, we have

$$\|\Sigma_{i=1}^n \lambda_i x_i\| = \vee_{i=1}^n |\lambda_i| \ \|x_i\|.$$

We say that an orthogonal set $D \subset E$ is orthonormal if $\|x\| = 1$ for all $x \in D$.

EXAMPLE 3.9. For $x \in Z_p$ and $n \in \{1,2,\ldots\}$, set

$$\binom{x}{n} = \frac{x(x-1)\cdots(x-n+1)}{n!} .$$

Set $\binom{x}{0} = 1$. Then the functions $\binom{\cdot}{0}$, $\binom{\cdot}{1}$, $\binom{\cdot}{2}$,... form an orthonormal basis (the Mahler basis) for $(C(Z_p \to Q_p)$, $\| \ \|_c)$.

EXAMPLE 3.10. Recall from §2 that we can write $x \in Z_p$ as $x = \Sigma_{k=0}^\infty b_k p^k$, where $b_k \in \{0,1,\ldots,p-1\}$ for all k. Given $m \in \{1,2,\ldots\}$, we write $m \lhd x$ if $m = \Sigma_{k=0}^N b_k p^k$ for some $N \in \mathbb{N}$ and adopt the convention that $0 \lhd x$ for all x. If $n \in \{1,2,\ldots\}$, then

$$\{m: m \triangleleft n,\ m \neq n\}$$

is finite and has a largest element (in the order defined by the relation $\triangleleft$) which we denote by $n\_$. The functions $e_0, e_1, \ldots$ defined by

$$e_n(x) = \begin{cases} 1, & \text{if } n \triangleleft x, \\ 0, & \text{otherwise,} \end{cases}$$

form an orthonormal basis (the van der Put basis) for $C(\mathbb{Z}_p \to \mathbb{Q}_p)$. If $f \in C(\mathbb{Z}_p \to \mathbb{Q}_p)$ has the expansion

$$f(x) = \Sigma_{n=0}^{\infty} a_n e_n(x),$$

then $a_0 = f(0)$ and $a_n = f(n) - f(n\_)$ for $n = 1, 2, \ldots$. Each of the functions e_n is locally constant; in fact, if $|x-y| < n^{-1}$, then $e_n(x) = e_n(y)$.

One of the interesting properties of the van der Put basis is that it is easy to read off the continuity properties of a function from the coefficients in its van der Put expansion. The following lemma should be compared with Lemma 63.1 in [Schikhof, 1984].

LEMMA 3.11. Suppose that $\varphi: \{p^{-n}\}_{n=0}^{\infty} \to [0,\infty[$ is a non-decreasing function. If $f \in C(\mathbb{Z}_p \to \mathbb{Q}_p)$ has the expansion

$$f(x) = \Sigma_{n=0}^{\infty} a_n e_n(x),$$

then

$$\lim_{\varepsilon \downarrow 0} \sup_{0<|x-y|\le\varepsilon} \frac{|f(x)-f(y)|}{\varphi(|x-y|)} = \lim\sup_{n\to\infty} \frac{|a_n|}{\varphi(|n-n\_|)}\ .$$

PROOF. Consider $x, y \in \mathbb{N}$ with $|x-y| = p^{-m} < 1$. If we write $x = \Sigma_{i=0}^{M} b_i p^i$ and $y = \Sigma_{i=0}^{N} c_i p^i$ with $b_i, c_i \in \{0,1,\ldots,p-1\}$

for all i, then $b_0 = c_0, \dots, b_{m-1} = c_{m-1}$, but $b_m \neq c_m$. Set $s_j = \Sigma_{i=0}^{j} b_i p^i$ for $j = m-1, \dots, M$ and set $t_k = \Sigma_{i=0}^{k} c_i p^i$ for $k = m-1, \dots, N$. Observe that $s_{m-1} = t_{m-1}$. Note also that $(s_j)_- = s_{j-1}$ for $j = m, \dots, M$ and $(t_k)_- = t_{k-1}$ for $k = m, \dots, N$. Applying the ultrametric inequality gives that

$$|f(x)-f(y)|$$
$$= (\vee_{j=m}^{M} |f(s_j)-f(s_{j-})|) \vee (\vee_{k=m}^{N} |f(t_k)-f(t_{k-})|).$$

Since $s_j \geq p^m$ for $j = m, \dots, M$, $t_k \geq p^m$ for $k = m, \dots, N$, and $|n-n_-| \leq p^{-m}$ for $n \geq p^m$, we see that

$$\frac{|f(x)-f(y)|}{\Phi(|x-y|)} \leq \sup \left\{ \frac{|f(n)-f(n_-)|}{\Phi(|n-n_-|)} : n \geq p^m \right\}$$
$$= \sup \left\{ \frac{|a_n|}{\Phi(|n-n_-|)} : n \geq p^m \right\}.$$

As $\mathbb{N}$ is dense in $\mathbb{Z}_p$ and f is continuous, it follows that the right-hand side in the statement is at least the left-hand side. The reverse inequality is obvious. ∎

4. Measurable Vector Spaces

DEFINITION 4.1. Suppose that E is a vector space (over K) and $\mathcal{B}$ is a σ-field of subsets of E. If the map

$$(x,y) \mapsto x+y, \quad x,y \in E,$$

is $\mathcal{B}\times\mathcal{B}/\mathcal{B}$ measurable, and if the map

$$(\alpha,x) \mapsto \alpha x, \quad \alpha \in K, \ x \in E,$$

is $\mathcal{B}(K)\times\mathcal{B}/\mathcal{B}$ measurable, then we say that the pair $(E,\mathcal{B})$ is a measurable vector space.

As in the Gaussian case, the framework of measurable vector space random variables enables us to give a unified treatment of the subject. In particular, by working in this degree of generality, we are able to treat processes as just random elements in an appropriate space (cf. [Fernique, 1975]).

The following definition and lemma introduce a broad class of measurable vector spaces which will turn out to be the ones which are of the most interest to us.

DEFINITION 4.2. Suppose that E is a vector space and F is a collection of linear functionals on E. If $\mathcal{B}$ is the σ-field generated by F, then we say that the triple $(E,F,\mathcal{B})$ satisfies the hypothesis (*).

LEMMA 4.3. Suppose that the triple $(E,F,\mathcal{B})$ satisfies the hypothesis (*), then $(E,\mathcal{B})$ is a measurable vector space.

PROOF. We first show that the map $(x,y) \mapsto x+y$ is $\mathcal{B}\times\mathcal{B}/\mathcal{B}$ measurable. Note first of all that

$$\mathcal{G} = \{C \in \mathcal{B}: \{(x,y): x+y \in C\} \in \mathcal{B}\times\mathcal{B}\}$$

is a σ-field. Suppose that $T_1,\ldots,T_n \in F$ and that $B_1,\ldots,B_n$ are Borel subsets of $\mathbb{R}$. Then

$$\{(x,y): x+y \in T_1^{-1}(B_1) \cap\cdots\cap T_n^{-1}(B_n)\}$$
$$= \cap_{i=1}^n\{(x,y): T_i(x)+T_i(y) \in B_i\}.$$

As the map $(x,y) \mapsto T_i(x)+T_i(y)$ is $\mathcal{B}\times\mathcal{B}$ measurable, we see that $\cap_{i=1}^n T_i^{-1}(B_i) \in \mathcal{G}$. Applying a standard monotone class theorem (see, for example, II.3 in [Williams, 1979]), we see that $\mathcal{G} = \mathcal{B}$, as required.

The proof that the map $(\alpha,x) \mapsto \alpha x$ is $\mathcal{B}(K)\times\mathcal{B}/\mathcal{B}$ measurable is similar. ∎

EXAMPLE 4.4. Let I be any index set. For each $i \in I$, define a linear functional $T_i: K^I \to K$ by setting $T_i((x_j)_{j\in I}) = x_i$. From Lemma 4.3, we see that if we let $\mathcal{B}$ be the σ-field generated by $\{T_i\}_{i\in I}$, then $(K^I,\mathcal{B})$ is a measurable vector space.

LEMMA 4.5. Suppose that $(E,\|\ \|_E)$ is a separable Banach space with dual E^*. If $\mathcal{B}$ is the Borel σ-field of E, then the triple $(E,E^*,\mathcal{B})$ satisfies the hypothesis (*).

PROOF. If suffices to show that the map $x \mapsto \|x\|_E$ is measurable with respect to the σ-field generated by E^*, but this follows by standard arguments from the ultrametric analogue of the Hahn-Banach theorem given in Appendix A.8 of [Schikhof, 1984] (cf. the proof of Proposition 7.1.1 in [Laha and Rohatgi, 1979]). ∎

LEMMA 4.6. Let $(E,F,\mathcal{B})$ satisfy the hypothesis (*). Let $(X_1,\ldots,X_n)$ and $(Y_1,\ldots,Y_n)$ be two n-tuples of E-valued random variables. Suppose that

$$\mathbb{E}\chi(\Sigma_i T_i(X_i)) = \mathbb{E}\chi(\Sigma_i T_i(Y_i))$$

for all n-tuples $(T_1,\ldots,T_n)$ drawn from the vector space of linear functionals spanned by F. Then $(X_1,\ldots,X_n)$ and $(Y_1,\ldots,Y_n)$ have the same law.

PROOF. To make our notation simpler, we may assume without loss of generality that the zero functional belongs to F. It then suffices to check for each finite collection $\{T_{ij}: 1 \le i \le n,\ 1 \le j \le m(i)\} \subset F$ that the $(\Sigma_i m(i))$-tuple $(T_{ij}(X_i))$ has the same law as $(T_{ij}(Y_j))$. Applying Lemma 2.3 in the $(\Sigma_i m(i))$-dimensional case, it therefore suffices to show that

$$\mathcal{E}\chi(\Sigma_{i,j}\alpha_{ij}T_{ij}(X_i)) = \mathcal{E}\chi(\Sigma_{i,j}\alpha_{ij}T_{ij}(Y_i))$$

for all $(\Sigma_i m(i))$-tuples (α_{ij}) drawn from K. But if we let $T_i = \Sigma_j\alpha_{ij}T_{ij}$, $1 \le i \le n$, then each T_i belongs to the vector space span F and the result follows by assumption. ■

5. Gaussian Random Variables

The following definition parallels the usual abstract definition in the Euclidean case (see, for example, [Fernique, 1975] or [Jain and Marcus, 1978]).

DEFINITION 5.1. Let $(E,\mathcal{B})$ be a measurable vector space and suppose that X is an E-valued random variable. We say that X is K-Gaussian if when X_1,X_2 are two independent copies of X and $(\alpha_{11},\alpha_{12})$, $(\alpha_{21},\alpha_{22}) \in K^2$ are orthonormal, then (X_1,X_2) has the same law as $(\alpha_{11}X_1+\alpha_{12}X_2,\alpha_{21}X_1+\alpha_{22}X_2)$.

Clearly, if $X = 0$ almost surely, then X is K-Gaussian. Our first order of business obviously should be to show that there are non-trivial K-Gaussian random variables.

THEOREM 5.2. A K-valued random variable X which is not almost surely 0 is K-Gaussian if and only if

$$P(X \in dx) = q^{-n}\Phi(q^{-n}|x|)\mu(dx)$$

for some $n \in \mathbf{Z}$ or, equivalently,

$$E\chi(\xi X) = \Phi(q^n|\xi|).$$

PROOF. Suppose that X has the given characteristic function for some $n \in \mathbf{Z}$. From (2.4) this is equivalent to X having the given law. If X_1, X_2 are two independent copies of X and $(\alpha_{11},\alpha_{12})$, $(\alpha_{21},\alpha_{22})$ is a pair of orthonormal vectors in K^2, then, recalling (2.5),

$$\begin{aligned}
&E\chi(\xi_1(\alpha_{11}X_1+\alpha_{12}X_2)+\xi_2(\alpha_{21}X_1+\alpha_{22}X_2)) \\
&= \Phi(q^n|\xi_1\alpha_{11}+\xi_2\alpha_{21}|)\Phi(q^n|\xi_1\alpha_{12}+\xi_2\alpha_{22}|) \\
&= \Phi(q^n|\xi_1\alpha_{11}+\xi_2\alpha_{21}|\vee|\xi_1\alpha_{12}+\xi_2\alpha_{22}|) \\
&= \Phi(q^n|\xi_1(\alpha_{11},\alpha_{21})+\xi_2(\alpha_{21},\alpha_{22})|) \\
&= \Phi(q^n|\xi_1|\vee|\xi_2|) \\
&= \Phi(q^n|\xi_1|)\Phi(q^n|\xi_2|) \\
&= E\chi(\xi_1X_1)E\chi(\xi_2X_2).
\end{aligned}$$

From Lemma 2.3, we have that $(\alpha_{11}X_1+\alpha_{12}X_2,\alpha_{21}X_1+\alpha_{22}X_2)$ has the same law as (X_1,X_2), and hence X is K-Gaussian.

Conversely, suppose that X is K-Gaussian. Put $\varphi(\xi) = \mathrm{Ex}(\xi X)$. Let X_1, X_2 be two independent copies of X. Since $(1,1),(0,1)$ is a pair of orthonormal vectors in K^2, we have that (X_1+X_2, X_2) has the same law as X_1, and so $(\varphi(\xi))^2 = \varphi(\xi)$. Thus $\varphi(\xi) \in \{0,1\}$ for all $\xi \in K$.

Suppose that $\xi_0 \neq 0$ with $\varphi(\xi_0) = 1$. Such a ξ_0 must exist since φ is continuous with $\varphi(0) = 1$. As $(1,\alpha),(0,1)$ is a pair of orthonormal vectors in K^2 for each $|\alpha| \le 1$, we see that $X_1+\alpha X_2$ has the same law as X_1. Thus $\varphi(\xi)\varphi(\alpha\xi) = \varphi(\xi)$ and, in particular, $\varphi(\alpha\xi_0) = 1$. This implies that $\varphi(\xi) = 1$ for all $|\xi| \le |\xi_0|$. Now $\varphi \not\equiv 1$, since X is not almost surely 0; so we must have that $\varphi(\xi) = \Phi(q^n|\xi|)$ for some $n \in \mathbb{Z}$. ■

COROLLARY 5.3. If X is a K-valued K-Gaussian random variable, then $X \in L^\infty(P)$. We have that

$$\mathrm{Ex}(\xi X) = \Phi(\|X\|_\infty |\xi|).$$

If X is not almost surely 0, then

$$P(X \in dx) = \|X\|_\infty^{-1}\Phi(\|X\|_\infty^{-1}|x|)\mu(dx).$$

PROOF. Clear. ■

EXAMPLE 5.4. With Theorem 5.2 in hand, we can construct examples of non-trivial K-Gaussian random variables on any non-trivial measurable vector space. If $\{e_1,\dots,e_n\}$ is a finite subset of non-zero elements of some measurable vector space $(E,\mathcal{B})$ and $X_1,\dots,X_n$ are non-trivial independent K-valued K-Gaussian random variables, then it is easy to see that $X = \Sigma_i X_i e_i$ is an E-valued K-Gaussian random variable. If we set $L = \{(x_1,\ldots,x_n) \in K^n: \Sigma_i x_i e_i = 0\}$, then L is a subspace with dimension at most (n-1). From the translation invariance of the Haar measure μ on K^n, it is easy to see that $\mu(L) = 0$. Since the law of $(X_1,\ldots,X_n)$ is absolutely continuous with respect to μ, we find that $X \neq 0$ almost surely.

6. A Zero-One Law

THEOREM 6.1. Suppose $(E,\mathcal{B})$ is a measurable vector space and D is a measurable subspace of E. If X is an E-valued K-Gaussian random variable, then $P(X \in D)$ is either 0 or 1.

PROOF. Let X_1,X_2 be two independent copies of X. For $n = 1,2,\ldots,$ set

$$A_n = \{X_1+\rho^n X_2 \in D,\ \rho^n X_1+X_2 \notin D\}.$$

Note first of all that the vectors $(1,\rho^n)$, $(\rho^n,1)$ are an orthonormal pair. Clearly, $|(1,\rho^n)| = |(\rho^n,1)| = 1$. From the ultrametric inequality and symmetry, we need only show that there is no $\lambda \in K$ such that

$$|(1,\rho^n)+\lambda(\rho^n,1)| = |1+\lambda\rho^n| \vee |\rho^n+\lambda| < 1.$$

If this was so, then $|1+\lambda\rho^n| < 1$, and we see from the isosceles triangle property that $|\lambda\rho^n| = 1$ and hence $|\lambda| = q^n$. Again applying the isosceles triangle property, this implies that $|\rho^n+\lambda| = q^n > 1$, which is a contradiction.

By definition we therefore have that $(X_1+\rho^n X_2, \rho^n X_1+X_2)$ has the same law as (X_1, X_2), and hence

$$P(A_n) = P(X \in D)\, P(X \notin D). \tag{6.1.1}$$

Observe also that if $m \neq n$, then the matrix

$$\begin{pmatrix} 1 & \rho^m \\ 1 & \rho^n \end{pmatrix}$$

is invertible. Thus, if both $X_1+\rho^m X_2$ and $X_1+\rho^n X_2$ belong to D, then both X_1 and X_2 belong to D, and hence $A_m \cap A_n = \emptyset$. It follows that

$$1 \geq P(\cup_n A_n) = \Sigma_n P(A_n),$$

and so from (6.1.1) we conclude that $P(X \in D) \in \{0,1\}$. ■

This result and its proof parallel Theoreme 1.2.1 in [Fernique, 1975]. For certain Gaussian processes, it is possible to generalise this type of result and obtain a similar zero-one dichotomy for the probability of belonging to a subgroup rather than a vector subspace (cf. [Kallianpur, 1970], [Jain, 1971] and [Cambanis and Rajput, 1973]). It is clear that such a result will not hold in general for our setting. For if X is a K-valued K-Gaussian random variable with $\|X\|_\infty > 0$, then $G = \{x: |x| \leq q^{-1}\|X\|_\infty\}$ is a subgroup of K and $0 < P(X \in G) = q^{-1} < 1$.

7. <u>Some Results Under the Hypothesis (*)</u>

THEOREM 7.1. Suppose that $(E,F,\mathcal{B})$ satisfies the hypothesis (*). If X is an E-valued random variable, then X is

K-Gaussian if and only if $T(X)$ is K-Gaussian for all $T \in \operatorname{span} F$.

PROOF. Suppose first of all that $T(X)$ is K-Gaussian for all $T \in \operatorname{span} F$. Let X_1, X_2 be two independent copies of X. Fix an orthonormal pair of vectors $(\alpha_{11}, \alpha_{12})$, $(\alpha_{21}, \alpha_{22})$ and a pair of functionals T_1, T_2 belonging to span F. From Lemma 4.6, in order to show that X is K-Gaussian, we need to check that

$$\begin{aligned} &E\chi(T_1(\alpha_{11}X_1+\alpha_{12}X_2)+T_2(\alpha_{21}X_1+\alpha_{22}X_2)) \\ &\quad = E\chi(T_1(X_1)+T_2(X_2)). \qquad (7.1.1) \end{aligned}$$

From Corollary 5.3 we see that the left-hand side of (7.1.1) is given by

$$\Phi(\|(\alpha_{11}T_1+\alpha_{21}T_2)X\|_\infty \vee \|(\alpha_{12}T_1+\alpha_{22}T_2)X\|_\infty).$$

Suppose that $\dim \operatorname{span}\{T_1(X), T_2(X)\} = 2$ (the cases where the dimension is 1 or 0 can be handled similarly and more easily). Let Y_1, Y_2 be an orthonormal basis for $\operatorname{span}\{T_1(X), T_2(X)\}$ in $L^\infty(P)$ (from Theorem 50.8 and Exercise 50.B of [Schikhof, 1984] such a basis exists) and write

$$T_1(X) = \beta_{11}Y_1+\beta_{12}Y_2,$$

$$T_2(X) = \beta_{21}Y_1+\beta_{22}Y_2.$$

We have

$$\begin{aligned} \|(\alpha_{11}T_1+\alpha_{21}T_2)X\|_\infty &= \|\alpha_{11}(\beta_{11}Y_1+\beta_{12}Y_2)+\alpha_{21}(\beta_{21}Y_1+\beta_{22}Y_2)\|_\infty \\ &= |\alpha_{11}\beta_{11}+\alpha_{21}\beta_{21}| \vee |\alpha_{11}\beta_{12}+\alpha_{21}\beta_{22}|. \end{aligned}$$

Similarly,

$$\|(\alpha_{12}T_1+\alpha_{22}T_2)X\|_\infty = |\alpha_{12}\beta_{11}+\alpha_{22}\beta_{21}| \vee |\alpha_{12}\beta_{12}+\alpha_{22}\beta_{22}|.$$

One can now readily check that the left-hand side of (7.1.1) is just

$$\Phi(|\beta_{11}(\alpha_{11},\alpha_{12})+\beta_{21}(\alpha_{21},\alpha_{22})| \vee |\beta_{12}(\alpha_{11},\alpha_{12})+\beta_{22}(\alpha_{21},\alpha_{22})|)$$
$$= \Phi((|\beta_{11}|\vee|\beta_{21}|) \vee (|\beta_{12}|\vee|\beta_{22}|))$$

by the orthonormality of $(\alpha_{11},\alpha_{12})$, $(\alpha_{21},\alpha_{22})$.

Also, the right-hand side of (7.1.1) is just

$$\Phi(\|\beta_{11}Y_1+\beta_{12}Y_2\|_\infty \vee \|\beta_{21}Y_1+\beta_{22}Y_2\|_\infty)$$
$$= \Phi((|\beta_{11}|\vee|\beta_{12}|) \vee (|\beta_{21}|\vee|\beta_{22}|)).$$

Therefore (7.1.1) holds and X is K-Gaussian.

The proof of the converse is straightforward and we omit it. ■

EXAMPLE 7.2. Let I be any index set. If $\{X_i\}_{i\in I}$ is any collection of K-valued random variables, then $X = \{X_i\}_{i\in I}$ is a K^I-valued random variable, where we equip K^I with the σ-field $\mathcal{B}$ generated by the family $\{T_i\}_{i\in I}$ of coordinate maps introduced in Example 4.4. It is clear from Theorem 7.1 that X is K-Gaussian if and only if for each finite subcollection $\{i_1,\ldots,i_n\} \subset I$ we have that

$$(T_{i_1}(X),\ldots,T_{i_n}(X)) = (X_{i_1},\ldots,X_{i_n})$$

is a K^n-valued K-Gaussian random variable. Following the usual Euclidean nomenclature we then say that X is a K-Gaussian process on I. For convenience we will often write X(i) for X_i.

COROLLARY 7.3. Suppose that $(E, \| \|_E)$ is a separable Banach space. If $\{X_n\}_{n=0}^{\infty}$ is a sequence of E-valued K-Gaussian random variable such that $X_n \to X$ almost surely as $n \to \infty$ for some random variable X, then X is also K-Gaussian.

PROOF. From Lemma 4.5 we have that $(E, E^*, \mathcal{B})$ satisfies the hypothesis (*), where $\mathcal{B}$ is the Borel σ-field of E. By Theorem 7.1, we therefore need only check that T(X) is K-Gaussian for all $T \in E^*$. However,

$$
\begin{aligned}
E\chi(\xi T(X)) &= E\chi(\lim_{n\to\infty} \xi T(X_n)) \\
&= \lim_{n\to\infty} E\chi(\xi T(X_n)) \\
&= \lim_{n\to\infty} \Phi(\|T(X_n)\|_\infty |\xi|) \\
&= \Phi(\|T(X)\|_\infty |\xi|),
\end{aligned}
$$

and the result follow from Corollary 5.3. ∎

If $(E, F, \mathcal{B})$ satisfies the hypothesis (*) and X is an E-valued random variable, then in order to show that X is K-Gaussian, we need only check that X satisfies a condition which is seemingly much weaker than the one given in Definition 5.1. The following result has no Euclidean counterpart.

COROLLARY 7.4. Suppose that $(E, F, \mathcal{B})$ satisfies the hypothesis (*). Let X be an E-valued random variable. Let X_1, X_2 be two independent copies of X. Then X is K-Gaussian if and only if $X_1 + \alpha X_2$ has the same law as X_1 for all $|\alpha| \le 1$.

PROOF. Suppose that $X_1+\alpha X_2$ has the same law as X_1 for all $|\alpha| \leq 1$. For any $T \in$ span F, we then have that $T(X_1)+\alpha T(X_2)$ has the same law as $T(X_1)$ for all $|\alpha| \leq 1$. A perusal of the proof of Theorem 5.2 shows that this is enough to imply that $T(X)$ is K-Gaussian. Applying Theorem 7.1, we see that X is K-Gaussian.

Conversely, given α with $|\alpha| \leq 1$, the pair $(1,\alpha)$, $(0,1)$ is orthonormal and so, by definition, $(X_1+\alpha X_2, X_2)$ has the same law as (X_1, X_2). ■

We finish this section with two results which further reinforce the connection between the properties of K-Gaussian random variables and the local field theory of orthogonality.

THEOREM 7.5. If $X = (X_1,\ldots,X_n)$ is a K^n-valued K-Gaussian random variable, then $\{X_1,\ldots,X_n\}$ is an orthogonal set in $L^\infty(P)$ if and only if $X_1,\ldots,X_n$ are independent.

PROOF. Suppose that $\{X_1,\ldots,X_n\}$ is an orthogonal set. From Theorem 7.1 we have that $\xi\cdot X$ is K-Gaussian for all $\xi = (\xi_1,\ldots,\xi_n) \in K^n$. Applying Corollary 5.3, we have

$$\begin{aligned} E\chi(\xi\cdot X) &= E\chi(1(\xi\cdot X)) \\ &= \Phi(\|\xi\cdot X\|_\infty) \\ &= \varphi(\|\Sigma_i \xi_i X_i\|_\infty) \\ &= \Phi(\vee_{i=1}^n |\xi_i| \; \|X_i\|_\infty) \end{aligned}$$

$$= \prod_{i=1}^{n} \Phi(\|X_i\|_\infty |\xi_i|)$$
$$= \prod_{i=1}^{n} E\chi(\xi_i X_i),$$

and the result follows from Lemma 2.3.

The proof of the converse is similar, and we omit it. ■

COROLLARY 7.6. Suppose that $X = (X_1,\dots,X_n)$ is a K^n-valued random variable. Then X is K-Gaussian if and only if for some m there exists a vector $Y = (Y_1,\dots,Y_m)$ of independent K-valued K-Gaussian random variables, and an m×n matrix A such that X = YA.

PROOF. Suppose that X = YA with Y and A as above. From Example 5.4, we observe that Y is K-Gaussian. From Theorem 7.1, we have for each $\xi \in K^n$ that $\xi\cdot X = (\xi A')\cdot Y$ is K-Gaussian and so, by Theorem 7.1, X is K-Gaussian.

Conversely, suppose that X is K-Gaussian. Let $\{Y_1,\dots,Y_m\}$ be an orthonormal basis for span$\{X_1,\dots,X_n\}$ in $L^\infty(P)$. Since $Y = (Y_1,\dots,Y_m) = XB$ for some n×m matrix B, we have by the argument of the previous paragraph that Y is K-Gaussian. From Theorem 7.5, $Y_1,\dots,Y_m$ are independent and the result follows. ■

THEOREM 7.7. Suppose that $X = (X_1,\dots,X_n)$ is a vector of independent identically distributed K-valued K-Gaussian random variables. Suppose that $\{\alpha_1,\dots,\alpha_m\}$ is an orthogonal set in K^n. Define an n×m matrix by

$A = (\alpha_1', \ldots, \alpha_m')$. Then $Y = XA$ is a vector of independent K-Gaussian random variables.

PROOF. For $\xi = (\xi_1, \ldots, \xi_m) \in K^m$ we have, setting $\sigma = \|X_1\|_\infty$, that

$$\begin{aligned} E\chi(\xi \cdot Y) &= E\chi(XA\xi') \\ &= E\chi(X \cdot (\xi A')) \\ &= \Pi_{j=1}^{n} \Phi(\sigma |(\xi A')_j|) \\ &= \Phi(\sigma \vee_{j=1}^{n} |(\xi A')_j|) \\ &= \Phi(\sigma |(\xi A')|) \\ &= \Phi(\sigma |\Sigma_{i=1}^{m} \xi_i \alpha_i|) \\ &= \Phi(\sigma \vee_{i=1}^{m} |\xi_i| \; |\alpha_i|) \\ &= \Pi_{i=1}^{m} \Phi((\sigma|\alpha_i|)|\xi_i|), \end{aligned}$$

and the result follows from Lemma 2.3 and Theorem 5.2. ■

8. Counterexamples

Apart from Corollary 7.4 our results have all had a Euclidean counterpart. In this section we give two examples of Euclidean results for which the natural local field analogue is false.

EXAMPLE 8.1. One of the main reasons why the theory of Gaussian processes is so tractable is that the law of such a process is completely determined by its family of 2-dimensional marginal distributions or, more precisely, by the mean and covariance structure of the process. There is, however, no fixed integer n such that the law of every K-Gaussian process is determined by its family of n-dimensional marginal distributions. Suppose that $\{Z_1,\ldots,Z_{n+1}\}$ is a set of independent identically distributed K-value K-Gaussian random variables with $\|Z_1\|_\infty = 1$. Set $X = (Z_1,\ldots,Z_n,Z_{n+1})$ and $Y = (Z_1,\ldots,Z_n,Z_1+\cdots+Z_n)$. One can check, using Theorem 7.7, that for each set of indices $\{i_1,\ldots,i_n\} \subset \{1,\ldots,n+1\}$ we have that $(X_{i_1},\ldots,X_{i_n})$ has the same law as $(Y_{i_1},\ldots,Y_{i_n})$. Clearly, however, X does not have the same law as Y.

EXAMPLE 8.2. A well-known theorem due to Cramér states that is X,Y are independent random variables and X+Y is centered Gaussian, then there is a constant c such that $X = Z+c$ and $X = W-c$, where both Z and W are centered Gaussian. In our setting, it is possible to choose two non-trivial independent K-valued random variables X,Y such that X+Y is K-Gaussian and yet neither X or Y has a law which is even absolutely continuous with respect to the Haar measure μ. To see this, let $\{a_1,\ldots,a_q\}$ be a complete set of coset representatives for $\{x: |x| \le q^{-1}\}$ in $\{x: |x| \le 1\}$. We may take $0 \in \{a_1,\ldots,a_q\}$. Set

$$\mu_0 = q^{-1}\Sigma_i\delta_i,$$

where δ_i is the unit point mass at a_i, and put

$$\mu_j(dx) = \mu_0(\rho^{-j}dx),\ j > 0.$$

It is easy to see that

$$\mu_0 * \mu_1 * \cdots * \mu_n(dx) \to \Phi(|x|)\mu(dx)$$

weakly as $n \to \infty$. If we put

$$\lambda_n = \mu_0 * \mu_2 * \cdots * \mu_{2n}$$

$$\nu_n = \mu_1 * \mu_3 * \cdots * \mu_{2n+1},$$

then there are probability measures λ and ν such that $\lambda_n \to \lambda$ weakly as $n \to \infty$, $\nu_n \to \nu$ weakly as $n \to \infty$, and $\mu = \lambda * \nu$. If we write $x \in \{y: |y| \le 1\}$ as $x = \sum_{k=0}^{\infty} x_k \rho^k$ where $x_k \in \{a_1,\ldots,a_q\}$, then

$$\text{supp}\ \lambda = \{x: 0 = x_1 = x_3 = \cdots\}$$

and

$$\text{supp}\ \nu = \{x: 0 = x_0 = x_2 = \cdots\},$$

so we have that

$$0 = \mu(\text{supp}\ \lambda) = \mu(\text{supp}\ \nu).$$

Therefore, if we construct two independent random variables X,Y with laws λ,ν, respectively, then the claim follows.

9. Random Series

Unlike the Gaussian case, the question of convergence of random series in our setting is almost trivial.

LEMMA 9.1. Suppose that $(E, \|\ \|_E)$ is a Banach space. Consider $\{f_n\}_{n=0}^{\infty} \subset E$ and a sequence $\{X_n\}_{n=0}^{\infty}$ of independent K-valued K-Gaussian random variables. The series $\sum_{n=0}^{\infty} X_n f_n$ converges almost surely in E if and only if $\|X_n\|_\infty \|f_n\|_E \to 0$ as $n \to \infty$.

PROOF. From the ultrametric inequality we have that the series converges if and only if $|X_n| \, \|f_n\|_E \to 0$ almost surely as $n \to \infty$. From the Borel-Cantelli lemmas we see that $|X_n| \, \|f_n\|_E \to 0$ if and only if

$$\Sigma_{n=0}^{\infty} P(|X_n| \, \|f_n\|_E > \varepsilon) < \infty \qquad (9.1.1)$$

for all $\varepsilon > 0$. Since $P(|X_n| \, \|f_n\|_E > \varepsilon) = 0$ when $\varepsilon \geq \|X_n\|_\infty \|f_n\|_E$ and $P(|X_n| \, \|f_n\|_E > \varepsilon) \geq 1-q^{-1}$ when $\varepsilon < \|X_n\|_\infty \|f_n\|_E$, it is clear that (9.1.1) occurs if and only if $\|X_n\|_\infty \|f_n\|_E > \varepsilon$ for finitely many n, and so the result follows. ∎

For the remainder of this section, we will be concerned with random series in the case when $K = \mathbb{Q}_p$ and our Banach space is $C(\mathbb{Z}_p \to \mathbb{Q}_p)$. In particular we will investigate the problem of representing stationary processes as random series. Our first result concerns the Mahler basis introduced in Section 3.

DEFINITION 9.2. Let $\{Z_n\}_{n=0}^{\infty}$ be a sequence of independent $\mathbb{Q}_p$-valued $\mathbb{Q}_p$-Gaussian random variables such that $\|Z_n\|_\infty = 1$ for all n. Suppose that $\{a_n\}_{n=0}^{\infty} \subset \mathbb{Q}_p$ is such that $|a_n| \to 0$ as $n \to \infty$. We say that the $\mathbb{Q}_p$-Gaussian processes X defined by $X(t) = \Sigma_{n=0}^{\infty} a_n Z_n \binom{t}{n}$, $t \in \mathbb{Z}_p$, is a random Mahler series.

Since $\|\binom{\cdot}{n}\|_C = 1$, we have from Example 5.4, Lemma 9.1, and Corollary 7.3 that X is a well-defined $C(\mathbb{Z}_p)$-valued $\mathbb{Q}_p$-Gaussian random variable and hence X is certainly a $\mathbb{Q}_p$-Gaussian process.

THEOREM 9.3. If $X = \{\Sigma_{n=0}^{\infty} a_n Z_n \binom{t}{n}\}$ is a random Mahler series, then X is stationary if and only if $|a_n| \geq |a_{n+1}|$ for all n.

PROOF. Set $Y(t) = X(t+1)$, $t \in \mathbb{Z}_p$. Since $\mathbb{N}$ is dense in $\mathbb{Z}_p$ and the paths of X are continuous, we see that X will be stationary if and only if the process Y has the same law as X. From the calculation on p. 152 of [Schikhof, 1984], we find that

$$Y(t) = \Sigma_{n=0}^{\infty}(a_n Z_n + a_{n+1} Z_{n+1})\binom{t}{n}.$$

It is clear that $\{a_n Z_n + a_{n+1} Z_{n+1}\}_{n=0}^{\infty}$ is a $\mathbb{Q}_p$-Gaussian process indexed by $\mathbb{N}$ and so X will be stationary if and only if

$$\|a_n Z_n + a_{n+1} Z_{n+1}\|_\infty = \|a_n Z_n\|_\infty \qquad (9.3.1)$$

for all n, and the sequence $\{a_n Z_n + a_{n+1} Z_{n+1}\}_{n=0}^{\infty}$ is independent. Observe from Theorem 7.5 that (9.3.1) is equivalent to requiring that

$$|a_n| \vee |a_{n+1}| = |a_n| \qquad (9.3.2)$$

Suppose that X is stationary, then (9.3.2) implies that $|a_n| \geq |a_{n+1}|$ for all n.

Conversely, suppose that $|a_n| \geq |a_{n+1}|$ for all n. Then (9.3.2) holds and applying Theorem 7.7, we see that the result will follow if we can show for each $n \in \mathbb{N}$ that the collection of vectors $\alpha_0 = (a_0, a_1, 0, \ldots, 0)$, $\alpha_1 = (0, a_1, a_2, 0, \ldots)$, ..., $\alpha_n = (0, \ldots, 0, a_n, a_{n+1})$ is orthogonal in $(\mathbb{Q}_p)^{n+1}$. However, for $\lambda_0, \ldots, \lambda_n \in \mathbb{Q}_p$, an induction based on Lemma 2.6 shows that

$$|\lambda_0\alpha_0+\cdots+\lambda_n\alpha_n|$$
$$= |\lambda_0|\ |a_0| \vee |\lambda_0+\lambda_1|\ |a_1| \vee\cdots\vee |\lambda_{n-1}+\lambda_n|\ |a_n| \vee |\lambda_n|\,|a_{n+1}|$$
$$= |\lambda_0|\ |a_0| \vee\cdots\vee |\lambda_n|\ |a_n|$$
$$= |\lambda_0|\ |\alpha_0| \vee\cdots\vee |\lambda_n|\ |\alpha_n|,$$

and so $\alpha_0,\ldots,\alpha_n$ are orthogonal, as required. ∎

Suppose that X is a stationary random Mahler series. If we define an isometric linear operator $L: C(\mathbf{Z}_p) \to C(\mathbf{Z}_p)$ by setting $(Lf)(x) = f(x+1)$, then, simply by the stationarity of X, we have that L^kX has the same law as X for $k = 0,1,\ldots$. The following result characterises what other "filters" of the form $b_0I+b_1L+\cdots+b_nL^n$ have this property.

THEOREM 9.4. Suppose that Q is a polynomial over $\mathbf{Q}_p$. The following are equivalent.

(i) For all stationary random Mahler series X, the law of $Q(L)X$ is the law of X.

(ii) The operator $Q(L)$ is an isometry of $C(\mathbf{Z}_p)$.

(iii) The polynomial Q has no roots in the set $\{x \in \mathbf{C}_p: |x-1| < 1\}$ and $|Q(1)| = 1$.

PROOF. We first show that (iii) ⇒ (i). Set $\Delta = L-I$ so that $(\Delta f)(x) = f(x+1)-f(x)$. If we define a polynomial R over $\mathbf{Q}_p$ by setting $R(x) = Q(1+x)$, then $Q(L) = R(\Delta)$ and the conditions of (iii) are equivalent to requiring that $|R(0)| = 1$ and R has no roots in the set $\{x \in \mathbf{C}_p: |x| < 1\}$.

If we write $R(x) = c_0+c_1x+\cdots+c_nx^n$, then we see from the discussion following Corollary 5.29 in [van Rooij, 1978] that these latter two conditions are in turn equivalent to requiring that

$$1 = |c_0| \geq \vee_{k=1}^{n} |c_k|. \tag{9.4.1}$$

Let $X(t) = \Sigma_{m=0}^{\infty} a_m Z_m \binom{t}{m}$ be a stationary random Mahler series. From the calculations in Section 5.2 of [Schikhof, 1984], we have that

$$(R(\Delta)X)(t) = \Sigma_{m=0}^{\infty} B_m \binom{t}{m},$$

where $B_m = \Sigma_{k=0}^{n} c_k a_{m+k} Z_{m+k}$. It is clear that $\{B_m\}_{m=0}^{\infty}$ is a $\mathbb{Q}_p$-Gaussian process indexed by $\mathbb{N}$ and that

$$\|B_m\|_\infty = \vee_{k=0}^{n} |c_k|\ |a_{m+k}| = |a_m|$$

(recall from Theorem 9.3 that $|a_0| \geq |a_1| \geq \cdots$). So, to establish that $R(\Delta)X$ has the same law as X, we need only show that elements of the sequence $\{B_m\}_{m=0}^{\infty}$ are independent or, equivalently, that B_m is independent of $\{B_{m+1}, B_{m+2}, \ldots\}$ for each $m \in \mathbb{N}$. For this, it certainly suffices to show that B_m is independent of $\{Z_{m+1}, Z_{m+2}, \ldots\}$ for each $m \in \mathbb{N}$. However, for any Borel set $G \subset \mathbb{Q}_p$, we have that

$$\begin{aligned} &P(B_m \in G \mid Z_{m+1}, Z_{m+2}, \ldots\} \\ &\quad = P(c_0 a_m Z_m \in G - \Sigma_{k=1}^{n} c_k a_{m+k} Z_{m+k} \mid Z_{m+1}, Z_{m+2}, \ldots) \\ &\quad = P(c_0 a_m Z_m \in G) \\ &\quad = P(B_m \in G), \end{aligned}$$

since

$$\begin{aligned}|\Sigma_{k=1}^{n} c_k a_{m+k} Z_{m+k}| &\le \|\Sigma_{k=1}^{n} c_k a_{m+k} Z_{m+k}\|_\infty \\ &= \vee_{k=1}^{n} |c_k|\ |a_{m+k}| \\ &\le |a_m|,\end{aligned}$$

and for any $\mathbb{Q}_p$-valued $\mathbb{Q}_p$-Gaussian random variable W we note that W+w has the same law as W, where w is any constant for which $|w| \le \|W\|$.

We now prove that (i) ⇒ (iii). In the notation above, if $R(\Delta)X$ has the same law as X, then, in particular, we must have that

$$|a_m| = \|B_m\|_\infty = \vee_{k=0}^{n} |c_k|\ |a_{m+k}|$$

for all $m \in \mathbb{N}$. By choosing $a_0 = 1$ and $0 = a_1 = a_2 = \cdots$, we see that $|c_0| = 1$, and by choosing $a_0 = a_1 = \cdots = a_n = 1$, we see that $|c_0| \ge \vee_{k=1}^{n} |c_k|$ (recall from Theorem 9.3 that such choices lead to stationary processes). Thus (9.4.1) and hence (iii) must hold.

Finally, the equivalence of (ii) and (9.4.1) is shown in the course of the proofs of Theorems 5.30 and 5.31 in [van Rooij, 1978]. ∎

A remarkable feature of the Gaussian theory is that stationary processes on the circle can be represented as random Fourier series with independent Fourier coefficients. With this in mind one might hope that all continuous stationary $\mathbb{Q}_p$-Gaussian processes on $\mathbb{Z}_p$ have the form given in Theorem 9.3. In Corollary 9.7 below, we show that not only is this not the case, but in fact there is no basis for $C(\mathbb{Z}_p)$ which "works." First, however, we obtain a result similar to Theorem 9.3 for the van der Put basis $\{e_n\}_{n=0}^{\infty}$ introduced in Section 3.

DEFINITION 9.5. Let $\{Z_n\}_{n=0}^{\infty}$ be a sequence of independent $\mathbb{Q}_p$-valued $\mathbb{Q}_p$-Gaussian random variables such that $\|Z_n\|_\infty = 1$ for all n. Suppose that $\{a_n\}_{n=0}^{\infty} \subset \mathbb{Q}_p$ is such that $|a_n| \to 0$ as $n \to \infty$. We say that the $\mathbb{Q}_p$-Gaussian process X defined by $X(t) = \Sigma_{n=0}^{\infty} a_n Z_n e_n(t)$, $t \in \mathbb{Z}_p$, is a random van der Put series.

As in the remarks following Definition 9.2, we see that X is indeed a well-defined $\mathbb{Q}_p$-Gaussian process with continuous sample paths.

THEOREM 9.6. If $X = \{\Sigma_{n=0}^{\infty} a_n Z_n e_n(t)\}$ is a random van der Put series, then X is stationary if and only if

$$|a_0| \geq |a_1| \geq |a_p| \geq \cdots \geq |a_{p^n}| \geq |a_{p^{n+1}}| \geq \cdots \qquad (9.6.1)$$

and

$$|a_{p^n}| = |a_{p^n+1}| = \cdots = |a_{p^{n+1}-1}| \qquad (9.6.2)$$

for all n.

PROOF. As in the proof of Theorem 9.3, we have that if we set $Y(t) = X(t+1)$, $t \in \mathbb{Z}_p$, then X will be stationary if and only if the process Y has the same law as X.

From Exercise 62.F in [Schikhof, 1984], we find that

$$Y(t) = \Sigma_{n=0}^{\infty} B_n e_n(t)$$

where

$$B_n = \begin{cases} a_0Z_0+a_1Z_1 & \text{if } n = 0, \\ a_{n+1}Z_{n+1}-a_{p^s}Z_{p^s} & \text{if } n = rp^s-1,\ s \in \mathbb{N},\ 2 \le r \le p, \\ a_{n+1}Z_{n+1} & \text{otherwise.} \end{cases}$$

It is clear that $\{B_n\}_{n=0}^{\infty}$ is a $\mathbb{Q}_p$-Gaussian process indexed by $\mathbb{N}$ and so X will be stationary if and only if

$$\|B_n\|_\infty = |a_n| \tag{9.6.3}$$

for all n and the sequence $\{B_n\}_{n=0}^{\infty}$ is independent.

Suppose that the conditions (9.6.1) and (9.6.2) hold. From Theorem 7.5, we have that (9.6.3) holds. Since $\{B_0,\dots,B_{p-1}\} \subset \operatorname{span}\{Z_0,\dots,Z_p\}$ and $\{B_{p^s},\dots,B_{p^{s+1}-1}\} \subset \operatorname{span}\{Z_{p^s+1},\dots,Z_{p^{s+1}}\}$ for $s = 1,2,\dots$, it therefore suffices by Theorem 7.5 to show that each of these subsets is orthogonal. From Theorem 7.5 and Lemma 2.7, we have

$$\begin{aligned} &\|\lambda_0B_0+\lambda_1B_1+\cdots+\lambda_{p-1}B_{p-1}\|_\infty \\ &\quad = \|\lambda_0(a_0Z_0+a_1Z_1)+\lambda_1(a_2Z_2-a_1z_1)+\cdots+\lambda_{p-1}(a_pZ_p-a_1Z_1)\|_\infty \\ &\quad = |\lambda_0|\ |a_0| \vee (|\lambda_0-(\lambda_1+\cdots+\lambda_{p-1})| \\ &\qquad\qquad \vee |\lambda_1|\vee\cdots\vee|\lambda_{p-1}|)|a_1| \vee |\lambda_{p-1}|\ |a_p| \\ &\quad = |\lambda_0|\ |a_0| \vee (|\lambda_1| \vee\cdots\vee |\lambda_{p-1}|)|a_1| \\ &\quad = |\lambda_0|\ \|B_0\|_\infty \vee |\lambda_1|\ \|B_1\|_\infty \vee\cdots\vee |\lambda_{p-1}|\ \|B_{p-1}\|_\infty, \end{aligned}$$

so that $B_0,\dots,B_{p-1}$ are orthogonal. A similar argument using Lemma 2.8 establishes that $B_{p^s},\dots,B_{p^{s+1}-1}$ are orthogonal for each $s = 1,2,\dots$ and hence completes the proof that X is stationary.

Conversely, if X is stationary, then by Theorem 7.5 the condition 9.6.3 is equivalent to requiring that

$$|a_0| = |a_0| \vee |a_1|,$$

$$|a_n| = |a_{n+1}| \vee |a_{p^s}| \text{ if } n = rp^{-s}-1, \quad s \in \mathbb{N},\ 2 \leq r \leq p,$$

and

$$|a_n| = |a_{n+1}| \text{ for all other } n.$$

It is straightforward to check that this implies (9.6.1) and (9.6.2). ■

COROLLARY 9.7. There is no linearly independent sequence of functions $\{f_n\}_{n=0}^{\infty} \subset C(\mathbb{Z}_p)$ such that every continuous stationary $\mathbb{Q}_p$-Gaussian process $X = \{X(t): t \in \mathbb{Z}_p\}$ is of the form $X(t) = \Sigma_{n=0}^{\infty} A_n f_n(t)$ for some sequence $\{A_n\}_{n=0}^{\infty}$ of independent $\mathbb{Q}_p$-valued $\mathbb{Q}_p$-Gaussian random variables.

PROOF. Suppose that $\{f_n\}_{n=0}^{\infty}$ has the requisite properties.

For $n \in \mathbb{N}$, set

$$X(t) = \Sigma_{k=0}^{p^n-1} B_k e_k(t), \quad t \in \mathbb{Z}_p,$$

where $\{e_k\}_{k=0}^{\infty}$ is the van der Put basis and $\{B_k\}_{k=0}^{p^n-1}$ is a set of independent $\mathbb{Q}_p$-Gaussian random variables with

$$\|B_0\|_{\infty} = \|B_1\|_{\infty} = \cdots = \|B_{p^n-1}\|_{\infty} = 1.$$

From Theorem 9.6, we see that X is stationary. By assumption we have that

$$X(t) = \Sigma_{k=0}^{\infty} A_k f_k(t),$$

where $(A_k)_{k=0}^{\infty}$ is a set of independent Q_p-Gaussian random variables.

By Theorem 7.5, the sequence $(A_k)_{k=0}^{\infty}$ is orthogonal, and hence the set $\{A_k: A_k \neq 0\}$ is linearly independent. A similar observation holds for $\{B_0,\ldots,B_{p^n-1}\}$. Since

$$\operatorname{span}(B_k) = \operatorname{span}(X(t): t \in \mathbf{Z}_p)$$
$$= \operatorname{span}(A_k),$$

we must have, in fact, that

$$X(t) = \Sigma_{k=0}^{p^n-1} A_{n,k} f_{n,k}(t)$$

where $\{f_{n,k}\}_{k=0}^{p^n-1} \subset \{f_k\}_{k=0}^{\infty}$ and $\{A_{n,k}\}_{k=0}^{p^n-1} = \{A_k: A_k \neq 0\}$.

Note also that we can write the support of the $C(\mathbf{Z}_p)$-valued random variable X as both

$$\left\{\Sigma_{k=0}^{p^n-1} b_k e_k: |b_0| \le 1,\ldots,|b_{p^n-1}| \le 1\right\}$$

and

$$\left\{\Sigma_{k=0}^{p^n-1} a_k f_{n,k}: |a_0| \le \|A_{n,0}\|,\ldots,|a_{p^n-1}| \le \|A_{n,p^n-1}\|\right\}.$$

From this we can deduce that

$$\operatorname{span}\{e_k\}_{k=0}^{p^n-1} = \operatorname{span}\{f_{n,k}\}_{k=0}^{p^n-1}.$$

Recalling that $\{e_k\}$ is a basis and that $\{f_k\}$ is linearly independent, we see that

$$\{f_k: k = 0,1,\ldots\} = \cup_{n=0}^{\infty}\{f_{n,k}: k = 0,\ldots,p^n-1\},$$

and also that each f_k is locally constant.

Now consider the process

$$Y(t) = C_0+C_1\binom{t}{1}, \quad t \in Z_p,$$

where C_0, C_1 are non-trivial independent $\mathbb{Q}_p$-Gaussian random variables with $\|C_0\|_\infty \geq \|C_1\|_\infty$. From Theorem 9.3 we have that Y is stationary. By arguments similar to those above, we find that

$$Y(t) = D_0f_0'+D_1f_1'$$

where $\{f_0', f_1'\} \subset \{f_k\}_{k=0}^\infty$ and D_0, D_1 are independent $\mathbb{Q}_p$-Gaussian random variables. But this is impossible since it would imply that, almost surely, the paths of Y are non-constant first-degree polynomials and also locally constant. ■

We finish with a result that gives the global and, in the stationary case, local continuity properties of a random van der Put series.

THEOREM 9.8. Suppose that $X = \{\Sigma_{n=0}^\infty a_n Z_n e_n(t)\}$ is a random van der Put series. If $\varphi: \{p^{-n}\}_{n=0}^\infty \to [0,\infty[$ is a non-decreasing function, then

$$\lim_{\varepsilon\downarrow 0} \sup_{0<|s-t|\leq\varepsilon} \frac{|X(s)-X(t)|}{\varphi(|s-t|)} = \lim\sup_{n\to\infty} \frac{|a_n|}{\varphi(|n-n\_|)}$$

almost surely. If, moreover, X is stationary, then the above lim sup coincides with

$$\lim_{\varepsilon\downarrow 0} \sup_{0<|t|\leq\varepsilon} \frac{|X(t)-X(0)|}{\varphi(|t|)}$$

almost surely.

PROOF. Given $\ell \in \mathbb{R}$, we see from Lemma 3.11 that

$$\lim_{\varepsilon\downarrow 0}\sup_{0<|s-t|\le\varepsilon} \frac{|X(s)-X(t)|}{\varphi(|s-t|)} \ge \ell$$

if and only if for all $\delta > 0$ we have that $|a_n Z_n| > (\ell-\delta)\varphi(|n-n\_|)$ for infinitely many n. Using a Borel-Cantelli argument similar to the one in Lemma 9.1, we see that this will be the case if and only if

$$\limsup_{n\to\infty}|a_n|/\varphi(|n-n\_|) \ge \ell,$$

and hence the first claim follows.

Suppose now that X is stationary. Since $(p^n)\_ = 0$, we have that

$$X(p^n)-X(0) = a_{p^n} Z_{p^n}.$$

From Theorem 9.6 we observe that $|a_{p^n}| = \cdots = |a_{p^{n+1}-1}|$, and from Lemma 53.3 of [Schikhof, 1984] we find that

$$p^{-n} = |p^n-p^n\_| = \cdots = |(p^{n+1}-1)-(p^{n+1}-1)\_|.$$

Again using a simple Borel-Cantelli argument, we have that

$$\begin{aligned}
&\lim_{\varepsilon\downarrow 0}\sup_{0<t\le\varepsilon}|X(t)-X(0)|/\varphi(|t|) \\
&\qquad \ge \limsup_{n\to\infty}|(X(p^n)-X(0)|/\varphi(p^{-n}) \\
&\qquad = \limsup_{n\to\infty}|a_{p^n} Z_{p^n}|/\varphi(p^{-n}) \\
&\qquad = \limsup_{n\to\infty}|a_{p^n}|/\varphi(p^{-n}) \\
&\qquad = \limsup_{n\to\infty}|a_n|/\varphi(|n-n_|),
\end{aligned}$$

and the second claim follows. ∎

10. Conclusion

There are obviously a multitude of analogies with the Gaussian theory which we have not explored in this paper.

Most noticeably, apart from Theorem 9.8, we have not had anything to say about sample path properties. It is possible to emulate the Gaussian construction of Proposition 4.1 in [Evans, 1986b] to obtain a broad class of stationary K-Gaussian processes indexed by $\{x \in K: |x| \leq 1\}$ with tractable distributional properties. In particular, one should be able to obtain results for this family on point hitting, Hausdorff dimensions, local times, multiple points, etc. similar to those given in [Evans, 1986a,b]. It would seem, however, to be very difficult to develop much of a theory in general, since we lack a convenient "handle" like the covariance to get an analytic hold on the distributional properties of a general process.

References

[1] S. Cambanis, B. S. Rajput (1973). Some zero-one laws for Gaussian processes. *Ann. Probab.* **1**, 304-312.

[2] S. N. Evans (1986a). Continuity properties of Gaussian stochastic processes indexed by a local field. *Proc. London Math. Soc.* to appear.

[3] S. N. Evans (1986b). Sample path properties of Gaussian stochastic processes indexed by a local field. *Proc. London Math. Soc.* to appear.

[4] X. Fernique (1975). *Lecture Notes in Mathematics*, no. 480. Springer.

[5] N. C. Jain (1971). A zero-one law for Gaussian processes. *Proc. Amer. Math. Soc.* **29**, 585-587.

[6] N. C. Jain, M. B. Marcus (1978). Continuity of sub-gaussian processes, in *Advances in Probability*, Vol. 4, Marcel Dekker.

[7] G. Kallianpur (1970). Zero-one laws for Gaussian processes. *Trans. Amer. Math. Soc.* **149**, 199-211.

[8] R. G. Laha, V. K. Rohatgi (1979). *Probability Theory*. Wiley.

[9] A. van Rooij (1978). *Non-archimedean Functional Analysis*. Marcel Dekker.

[10] W. H. Schikhof (1984). *Ultrametric Calculus*. Cambridge University Press.

[11] M. H. Taibleson (1975). *Fourier Analysis on Local Fields*. Princeton University Press.

[12] D. Williams (1979). *Diffusions, Markov Processes and Martingales*. Wiley.

Department of Mathematics
University of Virginia
Math-Astronomy Building
Charlottesville, VA 22903

SOME FORMULAS FOR THE ENERGY FUNCTIONAL OF A MARKOV PROCESS

by

P. J. Fitzsimmons* and R. K. Getoor*

1. Introduction

In this paper we shall establish two formulas relating the energy functional of a Markov process to that of a subprocess. Let X be a right Markov process and M an exact multiplicative functional of X. Writing (X, M) for the corresponding subprocess, let L and L^M denote the energy functionals of X and (X, M) respectively. Suppose that M doesn't vanish on $[0, \zeta[$, and define an additive functional A by $dA_t = -dM_t/M_{t-}$. Then given an X-excessive measure ξ and an X-excessive function u we have

$$L^M(\xi, u) = L(\xi, u) + \nu^\xi(u), \tag{1.1}$$

where ν^ξ is the Revuz measure of A relative to X and ξ. Formula (1.1) appears as (3.27) in **[GSt]** in the special case $M_t = e^{-qt}$.

The validity of (1.1) relies heavily on the strict positivity of M. Our second formula serves as a replacement for (1.1) in the general case. To state this formula we need the balayage operator R_M associated with $M; R_M$ operates on the cone of

* Research supported in part by NSF Grant DMS 87-21347.

X-excessive measures and is the dual of the operator P_M defined by

$$(1.2) \qquad P_M f(x) = \begin{cases} -P^x \int_0^\infty f\circ X_t \, dM_t, & x \in E_M \\ f(x), & x \in E\backslash E_M, \end{cases}$$

where $E_M = \{x \in E: P^x(M_0 = 1) = 1\}$. P_M is the analogue of the hitting operator P_B with which it coincides if $M = 1_{[0,T_B[}$, T_B being the hitting time of B. This notation established we have our second formula

$$(1.3) \qquad \begin{aligned} L(\xi, u) &= L^M(\tilde{\xi}, \tilde{u}) + L(\xi, P_M u) \\ &= L^M(\tilde{\xi}, \tilde{u}) + L(R_M \xi, u), \end{aligned}$$

where ξ and u are as for (1.1), $\tilde{\xi} = \xi - R_M \xi$, and $\tilde{u} = u - P_M u$. (See §3 for the precise definition of R_M, $\tilde{\xi}$, and $\tilde{u}$.)

Section 2 contains the proof of (1.1); this proof depends on (2.2) which is of interest in its own right. The balayage operator R_M is discussed in §3; here we omit many details, referring the reader to **[FG]** for full proofs in the special case $M = 1_{[0,T_B[}$. Formula (1.3) is proved in §4. The argument involves a probabilistic identification of the three terms in (1.3) and is perhaps of more interest than the formula itself.

In the remainder of this section we set down our notation and blanket hypotheses. Unexplained terminology can be found in **[Sh]** or **[FG]** and the reader is advised to have a copy of the latter at hand for reference.

Throughout the paper, $X = (\Omega, \mathcal{F}, \mathcal{F}_t, X_t, \theta_t, P^x)$ is a right Markov process in the sense of [**Sh**, §20] with state space $(E, \mathcal{E})$, semigroup (P_t), and resolvent (U^q). In particular, E is a separable Radon space and $\mathcal{E}$ is the σ-field of Borel sets for the Ray topology, $(\mathcal{F}_t)$ is the usual augmentation of the natural filtration of X, and conditions (20.4) and (20.5) in [**Sh**] are in force. The semigroup of X need only be subMarkovian, and ζ denotes the lifetime of X.

Let $S^q(X)$ denote the cone of q-excessive functions of X and put $\mathcal{E}^e = \sigma\{\bigcup_{q\geq 0} S^q(X)\}$. As a rule the letter f always denotes a positive $\mathcal{E}^e$-measurable function on E. The cone of q-excessive measures for X is denoted $\mathrm{Exc}^q(X)$. As usual, when $q = 0$ it vanishes from the notation; in particular $U = U^0$ is the potential kernel for X. The notational scheme for the various subcones of $\mathrm{Exc}(X)$ is as in [**FG**] and [**GSt**]. Thus $\mathrm{Pur}(X)$ (resp. $\mathrm{Inv}(X)$, $\mathrm{Dis}(X)$, $\mathrm{Con}(X)$) denotes the class of purely excessive (resp. invariant, dissipative, conservative) elements of $\mathrm{Exc}(X)$. The analogous classes over a subprocess (X, M) are denoted $S(X, M)$, $\mathrm{Exc}(X, M)$, $\mathrm{Pur}(X, M)$, etc.

We fix once and for all an exact multiplicative functional (MF), M, of X as specified in [**FG**, (2.1)]. As a matter of convention we assume that $M_t(\omega) = 0$ for all $t \geq \xi(\omega)$ and $\omega \in \Omega$. Define

(1.4) $$S = \inf\{t\colon M_t = 0\};$$

(1.5) $$E_M = \{x \in E\colon P^x(M_0) = 1\} = \{x \in E\colon P^x(S > 0) = 1\}.$$

Then S is a perfect, though not necessarily exact, terminal time and E_M, the set of permanent points of M, is $\mathcal{E}^e$-measurable and serves as the state space of the subprocess (X, M). We write (Q_t) and (V^q) for the semigroup and resolvent of (X, M). For example

$$Q_t f(x) = P^x(f\circ X_t M_t), \quad t \geq 0,\ x \in E.$$

Associated with M is the additive functional (AF) of (X, S)

$$dA_t = -1_{]0,S[}(t)\, dM_t / M_{t-}$$

which has Revuz measure (relative to ξ)

$$\nu^\xi(f) := \uparrow \lim_{t\downarrow 0} t^{-1} P^\xi \int_{]0,t]} f\circ X_s \, dA_s.$$

We write U^q_M for the q-potential operator of A; namely,

$$(1.6) \qquad U^q_M f(x) = -P^x \left(\int_{]0,S[} e^{-qt} f\circ X_t \, dM_t / M_{t-} \right).$$

The formula

$$(1.7) \qquad \nu^\xi(f) = \uparrow \lim_{q\to\infty} q\xi U^{q+p}_M f, \quad p \geq 0,$$

can be found in [**FG**, (2.15)]. Note that if $M_t = e^{-qt} 1_{[0,\zeta[}$, then $\nu^\xi = q\cdot\xi$. As noted by Meyer [**M66**], if $S \geq \zeta$ a.s., then

$$(1.8) \qquad U^q = V^q + U^q_M V^q;$$

this formula plays an important role in the sequel.

Finally, recall the energy functional $L\colon \mathrm{Exc}(X) \times S(X) \to [0,\infty]$ determined by

$$L(\xi, u) = \sup\{\mu(u) : \mu U \in \mathrm{Exc}(X),\ \mu U \leq \xi\},$$

for $\xi \in \mathrm{Exc}(X)$, $u \in S(X)$. The reader is referred to [**DM**, XII-39], [**GSt**], or [**FG**] for the various properties of the bilinear form L. The energy functional of (X, M) is denoted L^M, and that of the q-subprocess $X^q = (X, e^{-q\cdot} 1_{[0,\zeta[})$ is denoted L^q.

2. Proof of (1.1).

In this section we assume that the MF (M_t) satisfies $M_t > 0$ for all $0 \leq t < \zeta$ a.s.; that is, the terminal time S defined by (1.4) satisfies $P^x(S < \zeta) = 0$ for all $x \in E$. In particular, $E_M = E$.

We begin the proof of (1.1) by noting that it suffices to consider the two special cases (i) $\xi \in \mathrm{Dis}(X)$, and (ii) $\xi \in \mathrm{Con}(X)$. This is because all three terms in (1.1) are additive in ξ, and since each $\xi \in \mathrm{Exc}(X)$ admits a unique decomposition $\xi = \xi_d + \xi_c$ where $\xi_d \in \mathrm{Dis}(X)$, $\xi_c \in \mathrm{Con}(X)$.

(i) Assume $\xi \in \mathrm{Dis}(X)$ and choose a sequence (μ_n) of measures on E such that $\mu_n U \uparrow \xi$. Define $\nu_n = \mu_n + \mu_n U_M$, so that $\nu_n V = \mu_n U \uparrow \xi$ by (1.8) with $q = 0$. By well-known properties of the energy functionals L^M and L (see [**GSt**] or [**FG**]) we have

$$L^M(\xi, u) =\uparrow \lim \nu_n(u) =\uparrow \lim [\mu_n(u) + \mu_n U_M(u)]$$
$$= L(\xi, u) + \nu^\xi(u).$$

The last equality above follows from [**FG**, (2.18)]. (This argument is a trivial modification of that used in [**GSt**] for the special case $M_t = e^{-qt} 1_{[0,\zeta[}(\cdot)$)

Before proceeding with the proof of (1.1) in the case $\xi \in \mathrm{Con}(X)$ we record two facts

(2.1) LEMMA. *Given $\xi \in \mathrm{Con}(X)$ let $\xi = \xi_i + \xi_p$ be the decomposition of ξ into its invariant and purely excessive parts relative to (X, M). Then both ξ_i and ξ_p lie in $\mathrm{Con}(X)$.*

Proof. If $\xi(f) < \infty$ then since $\mathrm{Con}(X) \subset \mathrm{Inv}(X)$,

$$\xi_p P_t(f) = \xi P_t f - \xi_i P_t f \leq \xi(f) - \xi_i Q_t f$$
$$= \xi(f) - \xi_i(f) = \xi_p(f),$$

and so $\xi_p \in \mathrm{Exc}(X)$. But then $\xi_p \in \mathrm{Con}(X)$ since $\xi_p \leq \xi \in \mathrm{Con}(X)$. Therefore, $\xi_i = \xi - \xi_p$ is seen first to lie in $\mathrm{Inv}(X)$, then in $\mathrm{Con}(X)$, being dominated by ξ. ∎

Recall that (V^q) is the resolvent of (X, M) and that $V = V^0$.

(2.2) PROPOSITION. *(a) If $\xi \in \mathrm{Con}(X) \cap \mathrm{Inv}(X, M)$, then $\xi \in \mathrm{Con}(X, M)$, $\nu^\xi = 0$, and $M = 1_{[0,\zeta[}$ a.s. P^ξ.*

(b) *If $\xi \in \mathrm{Inv}(X)$, then $\xi_p = \nu^\xi V$, where ξ_p denotes the purely excessive part of ξ relative to (X, M).*

Before proving (2.2) let us use it and (2.1) to finish the proof of (1.1). Thus suppose that $\xi \in \mathrm{Con}(X)$. By Lemma (2.1) we need only consider the special cases $\xi \in \mathrm{Con}(X) \cap \mathrm{Inv}(X, M)$ and $\xi \in \mathrm{Con}(X) \cap \mathrm{Pur}(X, M)$. In the first case all terms in (1.1) vanish because of (2.2a) and [**GSt**, (3.11)]. In the second case $\xi = \nu^\xi V$ by (2.2b), hence $L^M(\xi, u) = \nu^\xi(u)$ while $L(\xi, u) = 0$ by the properties of L^M and L; see [**FG**] or [**GSt**]. ∎

Proof of (2.2). (a) Assume $\xi \in \mathrm{Con}(X) \cap \mathrm{Inv}(X, M)$. Then by (1.8), if $q > 0$,

$$\xi = q\xi U^q = q\xi V^q + q\xi U^q_M V^q = \xi + q\xi U^q_M V^q, \tag{2.4}$$

so $\xi U^q_M V^q = 0$ and $\xi U^q = \xi V^q$, even when $q = 0$. Since $U^q \geq V^q$, the last equality yields $U^q(x, \cdot) = V^q(x, \cdot)$ for ξ a.e. $x \in E$. But $\xi \in \mathrm{Con}(X)$ so if $f > 0$ then $Vf = Uf = \infty$ a.e. ξ, hence $\xi \in \mathrm{Con}(X, M)$. Moreover, $U^1 1 = V^1 1$ a.e. ξ, which implies that $M = 1_{[0,\zeta[}$ a.s. P^ξ. Since $V^q 1 > 0$ (recall that $E_M = E$), $\xi U^q_M V^q = 0$ implies $\xi U^q_M = 0$, hence $\nu^\xi = 0$ in view of (1.7).

(b) Now assume $\xi \in \mathrm{Inv}(X)$ and fix $f \in bp\mathcal{E} \cap L^1(\xi)$. Since $\xi \in \mathrm{Inv}(X) \subset \mathrm{Exc}(X, M)$, the first two equalities in (2.4) are valid; letting $q \to \infty$ we obtain

$$\lim_{q \to \infty} q\xi U^q_M V^q f = 0. \tag{2.5}$$

But by (1.7) and the resolvent equation for (V^q)

$$\nu^\xi V f = \uparrow \lim_{p \downarrow 0} \uparrow \lim_{q \uparrow \infty} q\xi U^q_M V^p f = \lim_{p \downarrow 0} \lim_{q \uparrow \infty} q\xi U^q_M (V^q f + (q-p) V^q V^p f).$$

Using now (2.5), (1.8), $\xi \in \mathrm{Inv}(X)$, and the resolvent equation, the last displayed line equals

$$\lim_{p \downarrow 0} \lim_{q \uparrow \infty} q(q-p)\xi(U^q - V^q)V^p f = \lim_{p \downarrow 0} \lim_{q \uparrow \infty} q\xi V^q f - p\xi V^p f = \xi_p(f),$$

because $p\xi V^p \to \xi_i$ as $p \to 0$. Thus $\nu^\xi V = \xi_p$ and the proof of (2.2) is complete. ∎

Remark. Suppose that $\xi \in \mathrm{Con}(X)$ is minimal; i.e., ξ admits no nontrivial decomposition into a sum of elements of $\mathrm{Exc}(X)$. Then by (2.1) and (2.2) either $\xi \in \mathrm{Inv}(X,M)$ in which case $M = 1_{[0,\zeta[}$ a.s. P^ξ, or $\xi = \nu^\xi V \in \mathrm{Pot}(X,M)$.

3. Balayage.

As in previous sections M is an exact MF of X, $S = \inf\{t: M_t = 0\}$, and $E_M = \{x \in E: P^x(M_0) = 1\}$. However we no longer assume that $S \geq \zeta$ a.s.

For $q \geq 0$ we define

$$(3.1) \qquad P_M^q f(x) = \begin{cases} -P^x \int_{]0,\zeta[} e^{-qt} f \circ X_t \, dM_t, & x \in E_M; \\ f(x), & x \in E_M^c, \end{cases}$$

which agrees with (1.2) when $q = 0$. Clearly $P_M^q(b\mathcal{E}^u) \subset b\mathcal{E}^u$ and $P_M^q(S^q(X)) \subset S^q(X)$. (Here and in the sequel, if $(F, \mathcal{F})$ is a measurable space, then $\mathcal{F}^u$ is the universal completion of $\mathcal{F}$.) As a replacement for (1.8) (which is valid only when $S \geq \zeta$), we have

$$(3.2) \qquad U^q = V^q + P_M^q U^q.$$

The operators P_M^q have duals relative to L^q which we now define following [**FG**, §3] (where the special case $M = 1_{[0,T_B[}$ was considered). First, if $q > 0$ and $\xi \in \mathrm{Exc}^q(X)$,

$$(3.3) \qquad R_M^q \xi(f) := L^q(\xi, P_M^q U^q f)$$

where L^q is the energy functional of the q-subprocess X^q. As in [**FG**], $R_M^q \xi \in \mathrm{Exc}^q(X)$, $R_M^q \xi \leq \xi$ and if $\mu_n U^q \uparrow \xi \in \mathrm{Exc}^q(X)$ then $R_M^q(\mu_n U) = \mu_n P_M^q U^q \uparrow R_M^q \xi$. Next, a straightforward computation shows that

$$(3.4) \qquad (q-r)U^r P_M^q U^q \leq P_M^r U^r - P_M^q U^q, \quad 0 < r < q.$$

Now if $\xi \in \text{Exc}^r(X)$ and $0 < r < q$, then we can choose a sequence of measures μ_n such that $\mu_n U^r \uparrow \xi$; setting $\nu_n = \mu_n + (q-r)\mu_n U^r$ we have $\nu_n U^q \uparrow \xi$, hence by (3.4),

$$R_M^q \xi = \lim_n \mu_n (I + (q-r)U^r) P_M^q U^q \leq \lim_n \mu_n P_M^r U^r = R_M^r \xi.$$

Thus $q \mapsto R_M^q \xi$ is decreasing on $]0,\infty[$ if $\xi \in \text{Exc}\,(X)$. We now define

$$R_M^0 \xi = R_M \xi = \uparrow \lim_{q \downarrow 0} R_M^q \xi, \quad \xi \in \text{Exc}\,(X). \tag{3.5}$$

Evidently $R_M\colon \text{Exc}\,(X) \to \text{Exc}\,(X)$. Various properties of R_M^q, $q \geq 0$, are established in [**FG**, §3] in the special case $M = 1_{[0,T_B[}$. These assertions remain valid here and will be used in the sequel. We mention here a few specifics. First, the relation (3.3) is true when $q = 0$ provided $\xi \in \text{Dis}\,(X)$. Second, the duality formula

$$L^q(R_M^q \xi, u) = L^q(\xi, P_M^q u), \tag{3.6}$$

is valid for $q \geq 0$, $\xi \in \text{Exc}^q(X)$, $u \in S^q(X)$. Finally, writing $\xi = \xi_d + \xi_c$ for the decomposition of $\xi \in \text{Exc}\,(X)$ into dissipative and conservative components, we have $(R_M \xi)_d = R_M(\xi_d)$ and $(R_M \xi)_c = R_M(\xi_c)$.

In the remainder of this section we give a precise definition of $\tilde{\xi}$ and $\tilde{u}$ appearing in formula (1.3). In fact we also consider the analogous objects $\tilde{\xi}^q$ and $\tilde{u}^q$, $q > 0$.

(3.7) PROPOSITION. *Fix $q \geq 0$. Given $\xi \in \text{Exc}^q(X)$, a σ-finite measure $\tilde{\xi}^q$ on $(E, \mathcal{E})$ is uniquely determined by*

$$\tilde{\xi}^q(f) = \xi(f) - R_M^q \xi(f), \quad f \in pL^1(\xi).$$

The measure $\tilde{\xi}^q$ is carried by E_M, and if regarded as a measure on $(E_M, \mathcal{E} \cap E_M)$ is an element of $\text{Exc}^q(X, M)$. If $\xi = \mu U^q$, then $\tilde{\xi}^q = \mu V^q$.

Proof. It suffices to consider the case $q > 0$. Then $\tilde{\xi}^q(E_M^c) = 0$ follows easily from (3.2) and (3.3), as does the evaluation $\tilde{\xi}^q = \mu V^q$ if $\xi = \mu U^q$. For a general $\xi \in \mathrm{Exc}^q(X)$ ($= \mathrm{Dis}^q(X)$ since $q > 0$), there is a sequence (μ_n) with $\mu_n U^q \uparrow \xi$. Then $\mu_n P_M^q U^q \uparrow R_M^q \xi$ as noted earlier; hence by (3.2), $\mu_n V^q f \to \tilde{\xi}^q(f)$ provided $f \in pL^1(\xi)$. Finally, for such f,

$$\tilde{\xi}^q Q_t^q f = \lim_n \mu_n V^q Q_t^q f \leq \lim_n \mu_n V^q f = \tilde{\xi}^q(f),$$

and so $\tilde{\xi}^q \in \mathrm{Exc}^q(X, M)$. ∎

The dual definition of $\tilde{u}^q$ is a more delicate matter.

(3.8) PROPOSITION. *Fix $q \geq 0$. Given $u \in S^q(X)$, there exists $\tilde{u}^q \in S^q(X, M)$ such that*

$$\tilde{u}^q = u - P_M^q u \quad \text{on} \quad \{P_M^q u < \infty\} \cap E_M.$$

If $u = U^q f$ then $\tilde{u}^q = V^q f$ on $\{P_M^q u < \infty\} \cap E_M$.

Proof. Suppose first that $u \in S^q(X)$ where $q > 0$. Let $B = \{P_M^q u < \infty\} \cap E_M$. Then $B \in \mathcal{E}^e$ is finely open. Since $\{P_M^q u < \infty\}$ is absorbing for X^q, each of the measures $Q_t^q(x, \cdot)$, $t \geq 0$, $x \in B$, is carried by B. Define u^* on E_M by

$$u^*(x) = \begin{cases} u(x) - P_M^q u(x), & x \in B; \\ \infty, & x \in E_M \backslash B. \end{cases}$$

We claim that u^* is (Q_t^q)-supermedian. Indeed since $q > 0$ there are functions f_n such that $U^q f_n \uparrow u$ as $n \to \infty$. Then $V^q f_n = U^q f_n - P_M^q U^q f_n \to u^*$ on B. Fatou's lemma now shows that $Q_t^q u^* \leq u^*$, $t \geq 0$, as claimed. Now define

$$\tilde{u}^q(x) = \uparrow \lim_{t \downarrow 0} Q_t^q u^*(x), \quad x \in E_M.$$

Clearly $\tilde{u}^q \in S^q(X,M)$. Writing $u_n^* = u^* \wedge n$ we see that each u_n^* is (Q_t^q)-supermedian and finely continuous on B. Thus if $x \in B$,

$$Q_t^q u_n^*(x) = P^x(u_n^*(X_t)e^{-qt}M_t) \to u_n^*(x), \quad \text{as} \quad t \downarrow 0. \tag{3.9}$$

But $u_n^* \uparrow u^*$ as $n \to \infty$, and $Q_t^q u_n^* \uparrow u_n^*$ on B as $t \downarrow 0$ because of (3.9). It follows that $\tilde{u}^q = u^*$ on B. In particular $\tilde{u}^q = u^* = V^q f$ on B if $u = U^q f$.

It remains to consider the case $q = 0$. Fix $u \in S(X) \subset S^q(X)$ and define $\tilde{u}^q$ as above. Note that $\{P_M u < \infty\} \subset \bigcap_{q>0} \{P_M^q < \infty\}$. Thus if $x \in \{P_M u < \infty\} \cap E_M$,

$$u^*(x) := u(x) - P_M u(x) = u(x) - \lim_{q \downarrow 0} P_M^q u(x)$$

(we put $u^* = \infty$ on $E_M \backslash \{P_M u < \infty\}$). Letting $q \downarrow 0$ in the inequality $Q_t^q \tilde{u}^q \leq \tilde{u}^q$ we see that u^* is (Q_t)-supermedian. Setting $\tilde{u} = \uparrow \lim_{t \downarrow 0} Q_t u^*$, the truncation argument used earlier shows that $\tilde{u} = u^*$ on $\{P_M u < \infty\} \cap E_M$. Moreover $\tilde{u} = Vf$ on $\{P_M u < \infty\} \cap E_M$ if $u = Uf$. ∎

Remark. It is easy to check that the mappings $\xi \mapsto \tilde{\xi}^q$ and $u \mapsto \tilde{u}^q$ are "positive linear" on their respective domains $\text{Exc}^q(X)$ and $S^q(X)$.

4. The Second Formula

We are now ready to give a precise statement, and proof, of formula (1.3). We shall first state and prove the result for a *Borel* right process X; i.e., E is a Lusin metrizable space with Borel sets $\mathcal{E}$, and each P_t maps $b\mathcal{E}$ into itself. This will enable us to use the Kuznetsov process associated with X and $\xi \in \text{Exc}(X)$, and also Meyer's perfection theorem **[M74]** for M. To the best of our knowledge neither the existence of the Kuznetsov process nor Meyer's theorem have been established for general right processes. In (4.19) we shall indicate how formula (3.1) for general right processes may be reduced to the Borel case. As in previous sections M is an exact MF of X. We maintain the notation established in previous sections.

(4.1) THEOREM. *Assume that X is a Borel right process. Given $\xi \in \mathrm{Exc}(X)$ and $u \in S(X)$,*

$$(4.2) \qquad L(\xi,u) = L^M(\tilde{\xi},\tilde{u}) + L(\xi, P_M u) = L^M(\tilde{\xi},\tilde{u}) + L(R_M\xi, u),$$

where $\tilde{\xi} \in \mathrm{Exc}(X,M)$ and $\tilde{u} \in S(X,M)$ are as defined in §3.

Of course, the second equality in (4.2) expresses the duality between P_M and R_M already noted. We leave it to the reader to check that these formulas are trivial if either $\xi = \mu U$ or $u = Uf$ provided the obvious finiteness conditions are satisfied. However we have been unable to use this fact to extend (4.2) to $\xi \in \mathrm{Dis}(X)$ in any straightforward manner. The problem is that $\xi \mapsto \tilde{\xi}$ is *not* monotone. .

Henceforth we assume that X is a Borel right process; see [**G75**] or [**Sh**, (20.6)]. Our notation for the Kuznetsov process (Y, Q_ξ) associated with X and $\xi \in \mathrm{Exc}(X)$ is that used in [**F**] and differs slightly from that used in [**FM**] or [**GSt**]. Let W denote the space of paths $w: \mathbb{R} \to E \cup \{\Delta\}$ that are E-valued and right continuous on an open interval $]\alpha(w), \beta(w)[$ and that take the value Δ elsewhere. The dead path $[\Delta]: t \to \Delta$ corresponds to $]\alpha, \beta[= \emptyset$, and the appropriate convention is $\alpha([\Delta]) = +\infty$, $\beta([\Delta]) = -\infty$. Let $Y = (Y_t : t \in \mathbb{R})$ denote the coordinate process on W, and put $\mathcal{G}^0 = \sigma\{Y_t : t \in \mathbb{R}\}$ and $\mathcal{G}^0_t = \sigma\{Y_s : s \le t\}$.

Given $\xi \in \mathrm{Exc}(X)$, the associated *Kuznetsov measure* Q_ξ is the unique measure on $(W, \mathcal{G}^0)$ not charging $\{[\Delta]\}$ such that if $t_1 < \cdots < t_n$,

$$(4.3) \qquad Q_\xi(\alpha < t_1, Y_{t_1} \in dx_1, \ldots, Y_{t_n} \in dx_n, t_n < \beta)$$
$$= \xi(dx_1) P_{t_2 - t_1}(dx_2) \cdots P_{t_n - t_{n-1}}(x_{n-1}, dx_n).$$

Evidently Q_ξ is σ-finite and *invariant* relative to the shift operators (σ_t) defined by $Y_t \circ \sigma_s = Y_{t+s}$, $s,t \in \mathbb{R}$. Also defined on W are birthing, killing, and shift operators:

$$\begin{aligned}(b_t w)(s) &= w(s) \quad , \quad s > t;\\ &= \Delta \quad , \quad s \le t;\\ (k_t w)(s) &= w(s) \quad , \quad s < t;\\ &= \Delta \quad , \quad s \ge t.\end{aligned}$$

$$\theta_t = b_0 \sigma_t = \sigma_t b_t, \quad t \in \mathbb{R}.$$

A convenient way to express the relationship between X and Y is to realize X on the "canonical" space $\Omega = \{w \in W : \alpha(w) = 0,\ Y_{\alpha+}(w) \text{ exists in } E\} \cup \{[\Delta]\}$. We let X_t, $\mathcal{F}_t^0$, and $\mathcal{F}^0$ denote the respective restrictions to Ω of Y_t, $\mathcal{G}_t^0$ and $\mathcal{G}^0$ if $t > 0$ and if $t = 0$ we let $X_0 = Y_{0+}|_\Omega$, $\mathcal{F}^0 = \mathcal{G}_{0+}^0|_\Omega$.

Clearly $X_{t+s} = X_s \circ \theta_t$ if $s,t \ge 0$, while $X_s \circ \theta_t = Y_{s+t}$ if $s \ge 0$, $\alpha < t < \infty$. See **[FM]**, **[GSt]**, **[G88]** and **[F]** for further details. Since X is Borel, a result of Meyer **[M74]** allows us to assume that the exact MF M is adapted to $(\mathcal{F}_{t+}^u)$. Following **[G88]** we now "extend" M to W: for $\alpha(w) < s \le t$ define

$$N(s,t) = \begin{cases} M_{t-s} \circ \theta_s, & \alpha < s < t;\\ 1, & \alpha < s = t.\end{cases} \tag{4.4}$$

This makes sense since $\theta_s : \{\alpha < s\} \to \Omega$. The map $s \mapsto N(s,t)$ is increasing and right continuous on $]\alpha, t[$. This allows us to define

$$N_t = \begin{cases} \downarrow \lim_{s \downarrow \alpha} N(s,t), & \alpha < t;\\ 1, & \alpha \ge t.\end{cases}$$

Clearly $t \mapsto N_t$ is decreasing and right continuous on $]\alpha, \infty[$, $N_t \circ \sigma_s = N_{t+s}$ for all $s,t \in \mathbb{R}$, and (N_t) is adapted to $(\mathcal{G}_{t+}^u)$. It follows from the multiplicative property of M that $N_s N(s,t) = N_t$ if $\alpha < s < t$. Letting $s \downarrow \alpha$, then $t \downarrow \alpha$, it follows that $N_{\alpha+}^2 = N_{\alpha+}$, whence

$$N_{\alpha+} = 0 \quad \text{or} \quad 1. \tag{4.5}$$

See [G88] for a complete discussion of these functionals.

The following evaluation is the key to Theorem (4.1).

(4.6) PROPOSITION. *If* $\xi \in \operatorname{Exc}(X)$, *then*

$$R_M\xi(f) = Q_\xi(f \circ Y_t(1 - N_t)), \quad \forall t \in \mathbb{R}. \tag{4.7}$$

Remarks. The R.H.S. of (4.7) is independent of $t \in \mathbb{R}$ owing to the (σ_t)-invariance of Q_ξ and the homogeneity of N_t. In [FM] the formula (4.7) was taken as the *definition* of R_M in the special case $M = 1_{[0,T_B[}$. The identification of this definition with that given in §3 of this paper was made in [FG].

Proof of (4.6). First suppose that $\xi \in \operatorname{Dis}(X)$. Define H on Ω by $H = \int_0^\infty f \circ X_t (1 - M_t)\, dt$ and note that H is $\mathcal{F}^u$-measurable and *excessive* in the sense that $s \mapsto H \circ \theta_s$ is decreasing and right continuous on $[0, \infty[$. Note that $P^\cdot(H) = P_M U f$. Define H^* on W by

$$H^* = \uparrow \lim_{s \downarrow \alpha} H \circ \theta_s.$$

It is shown in [F, (2.7)] that since $\xi \in \operatorname{Dis}$ there is a $\mathcal{G}^0$-measurable random time $S^* : W \to [-\infty, +\infty]$ such that $t + S^* \circ \sigma_t = S^*$ and $Q_\xi(S^* \notin \mathbb{R}) = 0$. Moreover, using [F, (4.4)],

$$R_M\xi(f) = L(\xi, P_M U f) = Q_\xi(H^*; 0 < S^* < 1). \tag{4.8}$$

(The first equality above is just (3.6).) But it is easy to check that

$$H^* = \int_{\mathbb{R}} f \circ Y_t (1 - N_t)\, dt = \int dt\, [f \circ Y_0 (1 - N_0)] \circ \sigma_t$$

so the third term in (4.8) is precisely $Q_\xi(f\circ Y_0(1-N_0))$ by the "switching identity" (2.1) in [F]. This proves (4.7) in case $\xi \in \mathrm{Dis}\,(X)$.

To handle $\xi \in \mathrm{Con}\,(X)$ we use the fact that $\mathrm{Exc}^q(X) = \mathrm{Dis}^q(X)$ if $q > 0$. Fix $\xi \in \mathrm{Con}\,(X)$ and let Q^q_ξ be the Kuznetsov process for X^q and ξ. Then as a special case of the results in [G88], if $F \in p\mathcal{G}^u$ with $F([\Delta]) = 0$,

$$Q^q_\xi(F) = Q_\xi \iint q^2 e^{-q(s-r)} 1_{\{r<s\}} F\circ k_s\circ b_r\, dr\, ds. \tag{4.9}$$

(One can verify (4.9) directly by checking finite dimensional distributions.) Now since $\xi \in \mathrm{Con}\,(X) \subset \mathrm{Dis}^q(X)$, by what has already been proved,

$$R^q_M\xi(f) = Q^q_\xi(f\circ Y_0(1-N_0)) = Q_\xi(f\circ Y_0 \int_{-\infty}^0 q\, e^{qr}(1-N_0\circ b_r)\, dr),$$

since $N_0\circ k_s = N_0$ if $\alpha < 0 < s$. If $r \le \alpha$ then $N_0\circ b_r = N_0$, while if $\alpha < r < 0$ then $N_0\circ b_r = M_{-r}\circ\theta_r$. Therefore

$$\begin{aligned} R^q_M\xi(f) = {} & Q_\xi(f\circ Y_0(1-N_0)e^{q\alpha}) \\ & + Q_\xi(f\circ Y_0 \int_\alpha^0 q e^{qr}(1-M_{-r}\circ\theta_r)\, dr). \end{aligned} \tag{4.10}$$

But $r \mapsto M_{-r}\circ\theta_r = N(r,0)$ is increasing and right continuous on $]\alpha, 0[$ with limit N_0 as $r \downarrow \alpha$. Integrating by parts in (4.10) we obtain

$$R^q_M\xi(f) = Q_\xi(f\circ Y_0 \int_{]\alpha,0]} e^{qr}\, d_r N(r,0)).$$

Consequently $R_M\xi(f) = \uparrow \lim_{q\downarrow 0} R^q_M\xi(f) = Q_\xi(f\circ Y_0(1-N_0))$, since $N(0,0) = 1$. The proof of Proposition (4.6) is complete. ■

For the proof of (4.2) in case $\xi \in \mathrm{Con}\,(X)$ we need the following extension of [GSt, (6.12)].

(4.11) COROLLARY. *Define* $\varphi = 1 - P^{\cdot}(M_{\zeta-})$. *Then* $\varphi \in S(X)$, *and if* $\xi \in$ Con(X) *then* $R_M\xi = \varphi \cdot \xi$.

Proof. The fact that $\varphi \in S(X)$ is a consequence of the exactness of M which implies that $M_{\zeta-}\circ\theta_t$ decreases to $M_{\zeta-}$ as t decreases to 0. Fix $\xi \in$ Con(X) so that $Q_\xi(\alpha > -\infty) = Q_\xi(\beta < +\infty) = 0$. Recall that (N_t) is decreasing and homogeneous: $N_t\circ\sigma_s = N_{t+s}$. Thus if $S_\lambda = \inf\{t > \alpha : N_t < \lambda\}$ then $t + S_\lambda\circ\sigma_t = S_\lambda$ for all $t \in \mathbb{R}$. Accordingly, by [F, (2.7)], since $\xi \in$ Con(X) we must have $Q_\xi(S_\lambda \in \mathbb{R}) = 0$ for all $\lambda \in]0,1[$. But $N_{\alpha+} = 0$ or 1 by (4.5); since $t \mapsto N_t$ is decreasing and $[0,1]$-valued, it follows that $N_t = N_{\alpha+}$ for all $t > \alpha$, a.s. Q_ξ. Consider now the "dual" $(\hat{N}_t)$ of (N_t) defined by

$$\hat{N}_t = \begin{cases} \uparrow \lim_{u\uparrow\beta} N(t,u) = M_{\zeta-}\circ\theta_t, & \alpha < t < \beta; \\ 0, & t \leq \alpha; \\ 1, & t \geq \beta. \end{cases}$$

Clearly $(\hat{N}_t)$ is increasing and homogeneous. Reasoning as for (N_t) we see that $\hat{N}_t = \hat{N}_{\beta-}$ for all $t < \beta$ a.s. Q_ξ. But $N(s,u) = N(s,t)N(t,u)$ for $\alpha < s < t < u < \beta$, so

$$N_{\beta-} = N_t\hat{N}_t = \hat{N}_{\alpha+}, \quad \alpha < t < \beta.$$

This and the fact that $Q_\xi(\alpha > -\infty) = Q_\xi(\beta < +\infty) = 0$ yields

$$N. \equiv \hat{N}. \equiv 0 \text{ or } 1 \text{ a.s. } Q_\xi.$$

But $\hat{N}_0 = M_{\zeta-}\circ\theta_0$ on $\{\alpha < 0 < \beta\}$, so using (4.7),

$$\begin{aligned} R_M\xi(f) &= Q_\xi(f\circ Y_0(1 - N_0)) = Q_\xi(f\circ Y_0(1 - \hat{N}_0)) \\ &= Q_\xi(f\circ Y_0(1 - M_{\zeta-}\circ\theta_0)) = Q_\xi(f\circ Y_0\, \varphi\circ Y_0) = \xi(\varphi \cdot f) \end{aligned}$$

as claimed. ∎

Proof of (4.2).

(i) $\xi \in \mathrm{Con}(X)$: Since $R_M\xi \in \mathrm{Con}(X)$ if $\xi \in \mathrm{Con}(X)$, it suffices to show that $\tilde{\xi} \in \mathrm{Con}(X, M)$, for then all terms in (4.2) will vanish. By (4.11), $\tilde{\xi} = (1-\varphi)\cdot\xi$. Since $P^{\xi}(\zeta < \infty) = 0$, if we write $M_\infty = \lim_{t\uparrow\infty} M_t$, then for $t \geq 0$, and ξ-a.e. $x \in E$,

$$\varphi(x) = P^x(1 - M_t) + P^x(M_t(1 - M_\infty \circ \theta_t))$$

$$= P^x(1 - M_t) + P^x(M_t \varphi \circ X_t) = P^x(1 - M_t) + \varphi(x) P^x(M_t),$$

where the last equality follows since $t \mapsto \varphi \circ X_t$ is constant a.s. P^{ξ} (see [**GSt**, (2.9)]). Letting $t \to \infty$ above we arrive at $\varphi = \varphi^2$ a.e. ξ. Thus $\tilde{\xi} = 1_{\{\varphi=0\}}\cdot\xi$. Choose $f \in pL^1(\xi)$ such that $\{f > 0\} = E_M$. Then $\xi\{0 < Uf < \infty\} = 0$ since $\xi \in \mathrm{Con}(X)$ (see [**B**] or [**D**]). Now $Uf = Vf + P_M Uf = Vf + \varphi \cdot Uf$ a.e. ξ since $t \mapsto Uf \circ X_t$ is constant a.s. P^{ξ} as noted earlier. Thus $Uf = Vf$ a.e. $\tilde{\xi}$ and so

$$\tilde{\xi}(0 < Vf < \infty) = \tilde{\xi}(0 < Uf < \infty) \leq \xi(0 < Uf < \infty) = 0.$$

Thus $\tilde{\xi} \in \mathrm{Con}(X, M)$ as desired.

(ii) $\xi \in \mathrm{Dis}(X)$: We first establish (4.2) in the special case $u = 1$, in which case $\tilde{1} = 1 - \varphi = \psi$ (say). Since $\xi \in \mathrm{Dis}(X)$, as noted earlier there is a random time $S^* \in \mathcal{G}^0$ such that $t + S^* \circ \sigma_t = S^*$ for all $t \in \mathbb{R}$, and $Q_\xi(S^* \notin \mathbb{R}) = 0$. We shall also need the sequence (S_n) of $(\mathcal{G}^0_{t+})$-stopping times constructed in [**FM**, (4.4)] so as to satisfy (a) $t + S_n \circ \sigma_t = S_n$ for all $t \in \mathbb{R}$, (b) $\alpha < S_n < \beta$ if $S_n < +\infty$, (c) $S_n \downarrow \alpha$ as $n \to \infty$ a.s. Q_ξ. As noted earlier, $N_{\beta-} = \hat{N}_{\alpha+} = \lim_{t\downarrow\alpha} M_{\zeta-}\circ\theta_t$. Thus by [**F**, (4.4)],

$$L(\xi, 1) = Q_\xi(0 < S^* < 1),$$

$$L(\xi, P_M 1) = Q_\xi(1 - N_{\beta-}; 0 < S^* < 1).$$

To prove (4.2) in the present case we must therefore show that

$$L^M(\tilde{\xi}, \tilde{1}) = Q_\xi(N_{\beta-}; 0 < S^* < 1). \tag{4.12}$$

To this end note that $N_{\beta-}$ is (σ_t)-invariant, so by [**F**, (2.4)] the R.H.S. of (4.12) may be written

$$\text{(4.13)} \qquad \lim_{n\to\infty} Q_\xi(N_{\beta-}; 0 < S^* < 1,\ S_n \in \mathbb{R}) = \lim_{n\to\infty} Q_\xi(N_{\beta-};\ 0 < S_n < 1),$$

since $Q_\xi(S^* \notin \mathbb{R}) = 0$. Now $\alpha < S_n < \beta$ a.s. Q_ξ on $\{S_n < 1\}$, so $N_{\beta-} = N_{S_n} M_{\zeta-} \circ \theta_{S_n}$ a.s. Q_ξ. The S_n being stopping times, we have

$$\text{(4.14)} \qquad Q_\xi(N_{\beta-};\ 0 < S_n < 1) = Q_\xi(N_{S_n} \psi \circ Y_{S_n};\ 0 < S_n < 1)$$

by the strong Markov property of (Y_t, Q_ξ). (Recall that $\psi = \tilde{1} = P^{\cdot}(M_{\zeta-})$.) Now if $t > S_n$ then $S_n \circ k_t = S_n$ and $Y_{S_n} \circ k_t = Y_{S_n}$. Also, by the construction of S_n in [**FM**], $\{S_n \circ k_t = +\infty\} = \{S_n \geq t\}$. Since $N_{S_n} = -\int_{]S_n,\infty]} dN_t$ $(N_\infty = 0)$, using (4.14) we obtain

$$Q_\xi(N_{\beta-};\ 0 < S_n < 1) = Q_\xi\left(-\int_{]\alpha,\infty]} (\psi \circ Y_{S_n}\ 1_{]0,1[}(S_n)) \circ k_t\, dN_t\right).$$

Theorem 4.12(iii) of [**T**] states that his last expression is

$$\text{(4.15)} \qquad Q_{\tilde{\xi}}^M(\psi \circ Y_{S_n};\ 0 < S_n < 1),$$

where $Q_{\tilde{\xi}}^M$ is the Kuznetsov measure for (X, M) and $\tilde{\xi} \in \operatorname{Exc}(X, M)$. (This may be verified by a comparison of finite dimensional distributions.) It is easy to check that $S_n \downarrow \alpha$ a.s. $Q_{\tilde{\xi}}^M$, so applying [**F**, (2.4)] as in (4.13), the expression in (4.15) tends to

$$L^M(\tilde{\xi}, \psi) = L^M(\tilde{\xi}, \tilde{1})$$

as $n \to \infty$. This combined with (4.12) and (4.13) yields (4.2) for $u = 1$.

In proving (4.2) for general $u \in S(X)$ we first consider the case $u < \infty$ a.e. ξ. We shall reduce this case to that previously considered by means of the u-transform of X. We refer the reader to **[GSt]** for a discussion of the relevant properties of u-transforms, and to [**Sh**, §62] for a complete discussion. According to [**GSh**, (6.19)], given $u \in S(X)$ there is a *Borel* measurable $\overline{u} \in S(X)$ such that $u = \overline{u}$ off an M-polar set. Each of the terms in (4.2) is unchanged if u is replaced by $\overline{u}$, so in the sequel we shall assume without loss of generality that u is Borel. The u-transform of X is denoted by $X^{(u)}$ and is the Borel right process on the state space $E_u = \{0 < u < \infty\}$ with semigroup $P_t^{(u)} f = u^{-1} P_t(uf)$. In general the superscript (u) will indicate objects defined relative to $X^{(u)}$. (Two exceptions are $P^{x/u}$, the law of $X^{(u)}$ started at $x \in E_u$, and L_u, the energy functional of $X^{(u)}$.)

The following result is well-known for hitting times (and for this result u need not be Borel).

(4.16) PROPOSITION. *Fix* $u \in S(X)$.

(a) $u \cdot P_M^{(u)} 1 = P_M u$ *on* $\{u < \infty\}$;

(b) *If* $\xi \in \mathrm{Exc}\,(X)$ *and* $\xi(u = \infty) = 0$, *then* $R_M^{(u)}(u \cdot \xi) = u \cdot R_M \xi$.

Proof. In this proof only we write K^q for the operator P_M^q taken relative to $X^{(u)}$ and W^q for the resolvent of $X^{(u)}$. As is well-known

$$u(x) P^{x/u}(F;\, t < \zeta) = P^x(F \cdot u \circ X_t), \quad F \in b\mathcal{F}_{t+}^u,$$

provided $u(x) < \infty$. In this case,

$$\begin{aligned} (4.17) \qquad u(x) K^q W^q f(x) &= u(x) P^{x/u} \int_0^\infty e^{-qt} f \circ X_t (1 - M_t)\, dt \\ &= P_M^q U^q(uf)(x). \end{aligned}$$

If $q > 0$ then there are bounded positive f_n such that $W^q f_n \uparrow 1$ on E_u. But then $U^q(f_n u) \uparrow u$ on $\{u < \infty\}$. Moreover, if $u(x) < \infty$ then $P_M^q(x, \{u = \infty\}) = 0$.

Replacing f by f_n in (4.17) and letting $n \to \infty$ we find that $u \cdot K^q 1 = P^q_M u$ on $\{u < \infty\}$. Passing to the limit as $q \downarrow 0$ establishes point (a). With the help of (4.7), point (b) follows exactly as in [**GSt**, (5.4ii)]. ∎

Now fix a (Borel measurable) $u \in S(X)$ with $\xi(u = \infty) = 0$. We apply (4.2) with $u = 1$, X replaced by $X^{(u)}$, and ξ replaced by $u \cdot \xi \in \mathrm{Dis}\,(X^{(u)})$:

$$(4.18) \qquad L_u(u\xi, 1) = (L_u)^M(\widetilde{u\xi}^{(u)}, \tilde{1}^{(u)}) + L_u(R^{(u)}_M(u\xi), 1).$$

But $L_u(u\xi, 1) = L(\xi, u)$ by [**GSt**, (4.10)]; combining this with (4.16b) shows that the third term in (4.18) reduces to $L(R_M\xi, u)$. It remains to show that the second term in (4.18) equals $L^M(\tilde{\xi}, \tilde{u})$. Put $u^* = u|_{E_M}$. Then $u^* \in S(X, M)$ and it is easy to check that $(X^{(u)}, M) = (X, M)^{(u^*)}$, hence $(L_u)^M = (L^M)_{u^*}$. By (4.16b),

$$\widetilde{u\xi}^{(u)} = u\xi - R^{(u)}_M(u\xi) = u\xi - uR_M\xi = u\tilde{\xi} = u^*\tilde{\xi},$$

since $\tilde{\xi}$ is carried by E_M. Also, by (4.16a) and the fact that $P^{(u)}_M 1 = 1$ off $E^{(u)}_M = E_M \cap E_u$,

$$u^*\tilde{1}^{(u)} = u - uP^{(u)}_M 1 = u - P_M u = \tilde{u}$$

on $E_M \cap \{u < \infty\}$, hence a.e. $\tilde{\xi}$. Combining these observations we see that

$$(L_u)^M(\widetilde{u\xi}^{(u)}, \tilde{1}^{(u)}) = L^M(\tilde{\xi}, \tilde{u}),$$

and (4.2) follows for $\xi \in \mathrm{Dis}\,(X)$ if $u < \infty$ a.e. ξ.

Finally, consider $\xi \in \mathrm{Dis}\,(X)$, $u \in S(X)$ and suppose that $\xi(u = \infty) > 0$. First note that $L(\xi, u) = \infty$. Indeed, choosing (μ_n) such that $\mu_n U \uparrow \xi$, we have $\mu_n U(u = \infty) > 0$ for all large n. But $\{u < \infty\}$ is absorbing so $U(x, \{u = \infty\}) = 0$ if $u(x) < \infty$. It follows that $\mu_n\{u = \infty\} > 0$ and that $L(\xi, u) = \uparrow \lim \mu_n(u) = \infty$.

Thus (4.2) will follow in the present case provided $\xi(P_M u = \infty) > 0$. By way of contradiction, assume that $\xi(P_M u = \infty) = 0$. Let $B = \{u = \infty\}$. Since $P_M u = u$ off E_M, $\xi(E_M \cap B) > 0$. Clearly $E_M \cap \{P_M u < \infty\} \subset \{u < \infty\}^r$ (the set of regular points for $\{u < \infty\}$), and since $\{u < \infty\}$ is absorbing, $E_M \cap \{P_M u < \infty\} \cap B^r = \emptyset$. Thus $\xi(E_M \cap B^r) = 0$. But $B^r \subset B$ and $\xi(B \backslash B^r) = 0$ ($B \backslash B^r$ is semipolar). Thus $\xi(E_M \cap B) = 0$ and we have our contradiction. The proof of (4.2) is at long last complete. ∎

(4.19) Remarks. We conclude with a brief indication of how Theorem (4.1) may be extended to general right processes. Roughly speaking, given $\xi \in \operatorname{Exc}(X)$, we produce a Borel right process X^* with the same finite dimensional distributions as X for ξ a.e. starting point. As far as formula (4.2) is concerned the processes X and X^* are "equivalent". Applying Theorem (4.1) to X^* we thereby obtain formula (4.2) for X.

Passing to the details, let $(\overline{X}_t, \overline{P}_t, \overline{E})$ be a Ray compactification of X as in [**Sh**, §39]. Fix $\xi \in \operatorname{Exc}(X)$ and choose $E_0 \in \overline{\mathcal{E}}$ such that $E_0 \subset E$ and $\xi(E \backslash E_0) = 0$. Let $E^* = D \cap \{x \in \overline{E}: \overline{U}^1(x, \overline{E} \backslash E_0) = 0\}$. Then $(E^*, \overline{\mathcal{E}} \cap E^*)$ is a Lusin space, $\xi(E \backslash E^*) = 0$, and E^* is absorbing for $\overline{X}$. It follows that X^*, the restriction of $\overline{X}$ to E^*, is a Borel right process and that $E^* \backslash E$ is quasi-polar for X^* (see [**Sh**, (39.15)]). In addition, $E \backslash E^*$ is ξ-polar for X.

If $\eta \in \operatorname{Exc}(X)$ and $\eta(E \backslash E^*) = 0$, then η may be regarded as a measure η^* on E^*; as such $\eta^* \in \operatorname{Exc}(X^*)$. Similarly, if $u \in S(X)$ then $u^*(x) := \uparrow \lim_{t \downarrow 0} \overline{P}_t u(x)$, $x \in E^*$, defines an element u^* of $S(X^*)$ such that $u^* = u$ on $E \cap E^*$. Writing L^* for the energy functional of X^*, we have $L^*(\eta^*, u^*) = L(\eta, u)$.

Now given an exact MF, M, of X, there exists an exact MF, M^*, of X^*, such that (using the obvious notation)

$$(P_M u)^* = P_{M^*}(u^*), \quad (R_M \eta)^* = R_{M^*}(\eta^*),$$

provided $u \in S(X)$, $\eta \in \mathrm{Exc}(X)$ with $\eta(E \backslash E^*) = 0$. We now apply Theorem (4.1) to X^* and the elements M^*, ξ^*, and u^*, and then verify that each term in (4.2) is unchanged if the $*$'s are dropped, whence (4.2) for X, M, ξ, and u. This task is routine, if lengthy, and is left to the interested reader.

References

[B] Blumenthal, R. M. (1986). A decomposition of excessive measures, in *Seminar on Stochastic Processes 1985*, pp. 1–8, Birkhäuser, Boston.

[D] Dynkin, E. B. (1980). Minimal excessive measures and functions, Trans. Amer. Math. Soc., **258**, 217–244.

[DM] Dellacherie, C., Meyer, P.-A. (1987). *Probabilités et Potentiel*, Ch. XII à XVI, Hermann, Paris.

[F] Fitzsimmons, P. J. (1988). On a connection between Kuznetsov processes and quasi-processes, in *Seminar on Stochastic Processes 1987*, pp.123-133, Birkhäuser, Boston.

[FG] Fitzsimmons, P. J., Getoor, R. K. (1988). Revuz measures and time changes. To appear in Math. Zeit.

[FM] Fitzsimmons, P. J. , Maisonneuve, B. (1986). Excessive measures and Markov processes with random birth and death, Probab. Th. Rel. Fields, **72**, 319-336.

[G75] Getoor, R. K. (1975). *Markov Processes: Ray Processes and Right Processes.* Lecture Notes in Math. **440**, Springer-Verlag, Berlin-Heidelberg-New York.

[G88] Getoor, R. K. (1988). Killing a Markov process under a stationary measure involves creation, Ann. Probab., **16**, 564-585.

[GSh] Getoor, R. K., Sharpe, M. J. (1984). Naturality, standardness, and weak duality for Markov processes, Z. Warscheinlichkeitstheorie verw. Geb., **67**, 1-62.

[GSt] Getoor, R. K. , Steffens J. (1987) The energy functional, balayage, and capacity, Ann. Inst Henri Poincaré, **23**, 321-357.

[M66] Meyer,P.-A. (1966). Quelques résultats sur les processus de Markov , Invent. Math. **1**, 101-115.

[M74] Meyer, P.-A. (1974). Ensembles aléatoires markoviens homogènes I, in *Séminaire de Probabilités VII*, Lecture Notes in Math., **321**, pp. 176-190, Springer-Verlag, Berlin-Heidelberg-New York.

[Sh] Sharpe, M. J. (1988). *The General Theory of Markov Processes*, Academic Press, New York.

[T] Toby, E. (1988). Birthing and killing a Markov process under a stationary measure, Ph. D. Thesis, University of California, San Diego.

P. J. Fitzsimmons
Department of Mathematics, C-012
University of California, San Diego
La Jolla, California 92093

R. K. Getoor
Department of Mathematics, C-012
University of California, San Diego
La Jolla, California 92093

Note on the 3G Theorem (d = 2)

Ira W. Herbst and Zhongxin Zhao

In this note, Theorem 2 in [1] is improved as follows.

__Theorem 2′__. (3G Theorem for d = 2) For a Jordan domain D in R^2 (see [1]), there exists a constant $C = C(D)$ such that

$$\frac{G_D(x,y)G_D(y,z)}{G_D(x,z)} \le C\,\frac{F(x,y)F(y,z)}{F(x,z)} \qquad x,y,z \in D, \tag{1}$$

where $F(x,y) = \max\left(\ell n \frac{1}{|x-y|}, 1\right)$. $F(x,y)$ can also be replaced by the Green function $G^k(x,y)$ of $\frac{\Delta}{2} - \frac{k^2}{2}$ in R^2 $(k > 0)$ since $G^k \sim F$ in $D\times D$.

__Proof of Theorem 2′__. (outline) By Theorem 1 in [1], we need only prove the following:

$$Q(x,y,z) \equiv \frac{\ell n\left[1+\frac{\delta(x)\delta(y)}{|x-y|^2}\right]\ \ell n\left[1+\frac{\delta(y)\delta(z)}{|y-z|^2}\right]}{\ell n\left[1+\frac{\delta(x)\delta(z)}{|x-z|^2}\right]} \le C\,F(x,y,z),$$

$$F(x,y,z) \equiv \frac{F(x,y)F(y,z)}{F(x,z)}, \qquad \delta(x) = \mathrm{dist}(x,\partial D) \tag{2}$$

We may assume the diameter of $D < \frac{1}{e}$, otherwise we can use the mapping $x \mapsto rx$ for a small $r > 0$ and the relation $G_D(x,y) = G_{rD}(rx,ry)$. Hence we have

$$F(x,y) = \ell n \frac{1}{|x-y|} \ge 1. \tag{3}$$

By symmetry, we may assume

$$|x-y| \le |y-z| \text{ and } |x-z| \le 2|y-z|. \tag{4}$$

Using the fact that $u\,\ell n \frac{1}{u}$ is increasing in $(0,\frac{1}{e}]$, we have

$$\frac{|x-z|^2}{|y-z|^2} \leq 4 \frac{\ell n \frac{1}{|y-z|^2}}{\ell n \frac{4}{|x-z|^2}} \leq C \frac{\ell n \frac{1}{|y-z|}}{\ell n \frac{1}{|x-z|}} . \qquad (5)$$

__Case (i)__. $\frac{\delta(x)\delta(z)}{|x-z|^2} < 5$ and $\delta(x) \leq |x-y|$: By (5), (3), $\delta(y) \leq \delta(x)+|x-y| < 2|x-y|$, and the fact that $\ell n(1+u) \geq cu$ in $(0,5]$, we have

$$Q(x,y,z) < C \frac{\delta(x)\delta(y)}{|x-y|^2} \cdot \frac{\delta(y)\delta(z)}{|y-z|^2} / \frac{\delta(x)\delta(z)}{|x-z|^2}$$

$$= C \frac{\delta(y)^2 |x-z|^2}{|x-y|^2 |y-z|^2} \leq C \frac{\ell n \frac{1}{|y-z|}}{\ell n \frac{1}{|x-z|}} \leq C F(x,y,z).$$

__Case (ii)__. $\frac{\delta(x)\delta(z)}{|x-z|^2} < 5$ and $\delta(x) > |x-y|$: By (5) and $\delta(y) \leq \delta(x)+|x-y| \leq 2\,\delta(x)$, we have

$$Q(x,y,z) \leq C \,\ell n \frac{1}{|x-y|} \cdot \frac{\delta(y)\delta(z)}{|y-z|^2} / \frac{\delta(x)\delta(z)}{|x-z|^2} \leq C F(x,y,z).$$

__Case (iii)__. $\frac{\delta(x)\delta(z)}{|x-z|^2} \geq 5$: Since $5 \leq \frac{\delta(z)[\delta(z)+|x-z|]}{|x-z|^2}$, we have $|x-z| \leq \delta(z)$. Similarly, $|x-z| \leq \delta(x)$. Hence, $\delta(z) \leq \delta(x)+|x-z| \leq 2\,\delta(x)$ and

$$\ell n\left[1+ \frac{\delta(x)\delta(z)}{|x-z|^2}\right] \geq \ell n\left[\frac{\delta(z)}{|x-z|}\right]. \qquad (6)$$

One can also prove

$$\ell n\left[1+ \frac{\delta(y)\delta(z)}{|y-z|^2}\right] \leq C \,\ell n\left[1+ \frac{\delta(z)}{|y-z|}\right]. \qquad (7)$$

Thus by using (6), (7) and the fact that for $0 < u \leq 1$, $\frac{\ell n(1+xu)}{\ell n(1+x)}$ is increasing with x in $(0,\infty)$, we have

$$Q(x,y,z) \leq C \,\ell n \frac{1}{|x-y|} \,\ell n\left[1+ \frac{\delta(z)}{|y-z|}\right] / \,\ell n\left[1+ \frac{\delta(z)}{|x-z|}\right]$$

$$\leq C \,\ell n \frac{1}{|x-y|} \,\ell n\left[1+ \frac{1}{|y-z|}\right] / \,\ell n\left[1+ \frac{1}{|x-z|}\right] \leq CF(x,y,z). \qquad \blacksquare$$

[1] Z. Zhao, Green functions and conditioned gauge theorem for a 2-dimensional domain, Seminar on Stochastic Processes, 1987.

THE INDEPENDENCE OF HITTING TIMES AND HITTING POSITIONS TO SPHERES FOR DRIFTED BROWNIAN MOTIONS

H. R. HUGHES and M. LIAO*

A drifted Brownian motion X_t is a diffusion process on $\mathbf{R}^n$ whose infinitesimal generator has the form

$$L = \frac{1}{2}\Delta + b,$$

where Δ is the usual Laplace operator and

$$b = \sum_{i=1}^{n} b^i(x) \frac{\partial}{\partial x^i}$$

is a smooth vector field on $\mathbf{R}^n$. When $b \equiv 0$, X_t becomes the usual n-dimensional Brownian motion.

Fix $r > 0$ and let

$$T = \inf\{t > 0;\ d(X_0, X_t) = r\},$$

the hitting time of the r-sphere centered at the starting point X_0. (T depends on the radius, r, but to simplify notation we do not indicate the radius.) It is well known that, for Brownian motion, T and X_T are independent under P^x, the probability measure associated with the process starting from x in $\mathbf{R}^n$; i.e., for any bounded Borel function ϕ on $[0,\infty)$ and any bounded Borel function ψ on $S_r(x)$, the r-sphere around x,

(1) $$E^x[\phi(T)\psi(X_T)] = E^x[\phi(T)]E^x[\psi(X_T)].$$

* Research supported in part by the Natural Sciences Foundation of P. R. China

A drifted Brownian motion, in general, does not have this independence property. However, for suitably chosen b, (1) may still hold for any x in $\mathbf{R}^n$ and r > 0. For example, when b is a constant vector field, as observed in [C]. The purpose of this paper is to try to determine the class of vector fields b such that the corresponding processes X_t have this independence property. We are unable to solve this problem completely, but in the case where b is a gradient vector field, we obtain a complete characterization. This result leads us to speculate on the possibility that independence implies $b = \nabla f$ for some function f. However, we find an example of a non-gradient vector field b for which the corresponding process X_t has this independence property.

See [H], [KO] and [L] for discussions of the related independence property on Riemannian manifolds.

Our result will also apply to a drifted Brownian motion in an open subset D of $\mathbf{R}^n$, killed upon reaching the boundary. Then we will only be concerned with the independence of the hitting time and the hitting position to any sphere entirely contained in D. Let h be a positive harmonic function in D and X_t be the h-transform of the Brownian motion in D. See Chapter X of [D] for the definition of the h-transform of a Brownian motion. X_t is a drifted Brownian motion with drift vector field $b = \nabla \log h$. We can check directly that X_t has the desired independence property. Note that when $D = \mathbf{R}^n$, h must be a constant, so our example is non-trivial only when $D \neq \mathbf{R}^n$. Independence for this case as well as the case of constant drift will follow directly from a more general result.

Before we state our result, let us first observe that, by Cameron-Martin-Girsanov (e.g., see VI.6 in [IW]), the independence

(1) is equivalent to the following statement. For any bounded Borel functions ϕ on $[0,\infty)$ and ψ on $S_r(x)$,

(2) $$E[\phi(T')\psi(x + W_{T'})M_{T'}] = E[\phi(T')M_{T'}]E[\psi(x + W_{T'})M_{T'}],$$

where $W_t = (W_t^1,\ldots,W_t^n)$ is an n-dimensional Brownian motion starting from the origin o, T' is the hitting time to $S_r = S_r(o)$ for W and M_t is the exponential martingale defined by

(3) $$M_t = \exp\{\int_0^t b(x + W_s)\circ dW_s - \frac{1}{2}\int_0^t [\nabla\cdot b + |b|^2](x + W_s)ds\},$$

where $\circ dW_s$ denotes the Stratonovich stochastic differential.

Now assume $b = \nabla f$ for some function f which satisfies

(4) $$\Delta f + |\nabla f|^2 = \text{constant}.$$

Let C be the above constant. Then $\nabla\cdot b + |b|^2 = C$ and

$$M_{T'} = \exp\{\int_0^{T'} \nabla f(x + W_s)\circ dW_s - \frac{1}{2}\int_0^{T'} [\Delta f + |\nabla f|^2](x + W_s)ds\}$$

$$= \exp\{f(x + W_{T'}) - f(x) - \tfrac{1}{2}CT'\}$$

$$= \exp\{f(x + W_{T'}) - f(x)\}\exp\{-\tfrac{1}{2}CT'\}.$$

From this and the independence of T' and $W_{T'}$, and the fact that $E[M_{T'}] = 1$, we can prove (2).

Therefore, when $b = \nabla f$, (4) is a sufficient condition for independence. In fact, it is also a necessary condition and we have the following result.

PROPOSITION: Assume $b = \nabla f$ for some function f. Then X_t has the independence property (1), for any x in $\mathbf{R}^n$ and $r > 0$, if and only if (4) holds.

To prove the above proposition, it remains to show that the independence property (1) implies (4). This will follow directly from the lemma below by setting $b = \nabla f$.

Lemma: Fix x_0 in $\mathbf{R}^n$ and let g be any smooth function defined on $\mathbf{R}^n - \{x_0\}$ such that g is constant along any ray starting from x_0, then for small $r > 0$,

$$E^{x_0}[Tg(X_T)] - E^{x_0}[T]E^{x_0}[g(X_T)]$$

$$(5) \quad = \lambda r^4 \sum_{i,h} \left[\frac{\partial}{\partial x^h}\frac{\partial}{\partial x^h}b^i - 6b^h\frac{\partial}{\partial x^i}b^h - 4\frac{\partial}{\partial x^i}\frac{\partial}{\partial x^h}b^h\right](x_0)\, I_r(x^i g)(x_0)$$

$$+ O(r^6),$$

where λ is a positive constant which depends only on the dimension n and $I_r(x^i g)(x_0)$ denotes the average of the function $x^i g(x)$ over the sphere $S_r(x_0)$.

The proof of the above lemma, which is given in [H] (in fact, a more general situation is considered there), is too complicated to be given completely here. We will only indicate the main ingredients of the proof.

Without loss of generality, we may assume $x_0 = o$. For Brownian motion, $E^o[T^k] = O(r^{2k})$. By the formula of Cameron-Martin-Girsanov, this holds also for a drifted Brownian motion. (In fact, this holds for any non-degenerate diffusion. See [GP].) Let D_r be the open disk of radius r centered at o. Recall $L = \frac{1}{2}\Delta + b$. For small $r > 0$, suppose we can find functions U_r and V_r such that

(6) $\quad LU_r = O(r^2)$ in D_r and $U_r = g$ on S_r,

(7) $\quad LV_r + U_r = O(r^4)$, $L^2V_r = O(r^2)$ in D_r, and $V_r = 0$ on S_r.

By Dynkin's formula and (6), we have

$$E^o[g(X_T)] = E^o[U_r(X_T)] = U_r(o) + E^o[\int_0^T LU_r(X_t)dt].$$

Since $LU_r(X_t) = O(r^2)$ for $t < T$ and $E^o[T] = O(r^2)$, we have

(8) $$E^o[g(X_T)] = U_r(o) + O(r^4).$$

By Dynkin's formula and Ito's formula for stochastic integrals, we obtain the following stochastic Taylor's formula.

$$E^o[V_r(X_T)] = V_r(o) + E^o[T\ LV_r(X_T)] - E^o[\int_0^T tL^2V_r(X_t)dt].$$

By (6) and (7), we have

(9) $$E^o[Tg(X_T)] = V_r(o) + O(r^6).$$

Setting $g = 1$, we obtain an estimate for $E^o[T]$. Therefore, in order to prove (5), we need only to find functions U_r and V_r satisfying (6) and (7), and to compute $U_r(o)$ and $V_r(o)$.

Define u_0 and v_0 by solving the following Dirichlet problems.

$$\tfrac{1}{2}\Delta u_0 = 0 \text{ in } D_1 \text{ and } u_0 = g \text{ on } S_1,$$

$$\tfrac{1}{2}\Delta v_0 + u_0 = 0 \text{ in } D_1 \text{ and } v_0 = 0 \text{ on } S_1$$

Let

$$L_1 = \sum_{i=1}^n b^i(o) \frac{\partial}{\partial x^i}$$

and for $k = 2,3,\ldots$, let

$$L_k = \frac{1}{(k-1)!} \sum_{i,i_1,\ldots,i_{k-1}} x^{i_1}\ldots x^{i_{k-1}} \frac{\partial}{\partial x^{i_1}}\cdots\frac{\partial}{\partial x^{i_{k-1}}} b^i(o) \frac{\partial}{\partial x^i}.$$

Define u_k and v_k inductively by the following recursive formulas.

$$\tfrac{1}{2}\Delta u_k + \sum_{j=1}^k L_j u_{k-j} = 0 \text{ in } D_1 \text{ and } u_k = 0 \text{ on } S_1,$$

$$\tfrac{1}{2}\Delta v_k + \sum_{j=1}^k L_j v_{k-j} + u_k = 0 \text{ in } D_1 \text{ and } v_k = 0 \text{ on } S_1.$$

If we let

$$U_r(x) = \sum_{k=0}^{3} u_k(\frac{x}{r})r^k$$

and

$$V_r(x) = \sum_{k=0}^{3} v_k(\frac{x}{r})r^{k+2},$$

then (6) and (7) hold. Therefore, in order to prove (5), we need only to compute $u_k(o)$ and $v_k(o)$ for $k = 1,2,3$. u_0 is given by Poisson's formula. The values of the other functions at o can be determined using the following Pizetti's formula (see [CH]). If the function u satisfies $\Delta^{k+1}u = 0$ for some integer $k \geq 0$, then

$$I_r(u)(o) = u(o) + \sum_{h=1}^{k} \frac{1}{2^h h! n(n+2)\cdots(n+2h-2)} \Delta^h u(o) r^{2h}.$$

The reader is referred to [H] for the details.

EXAMPLE: Finally, we give an example to show that when the drift vector field b is not a gradient, T and X_T may still be independent. We will consider the 2-dimensional case. Let p be a vector field on $\mathbf{R}^2$ defined by

$$p = -x^2\frac{\partial}{\partial x^1} + x^1\frac{\partial}{\partial x^2}.$$

Let $b = \nabla f + p$, where f is a function which is chosen so that

$$\nabla \cdot b + |b|^2 = \text{constant}.$$

Such a function f exists, at least locally. Let C be the above constant and let M_t be the exponential martingale given in (3). Then

$$M_{T'} = N_1 N_2,$$

where

$$N_1 = \exp\{f(x + W_{T'}) - f(x) + x^1 W_{T'}^2 - x^2 W_{T'}^1\}$$

and

$$N_2 = \exp\{-\tfrac{1}{2}CT' + \int_0^{T'} W_t^1 \circ dW_t^2 - \int_0^{T'} W_t^2 \circ dW_t^1\}.$$

In order to prove (2), it suffices to show that N_1 and N_2 are independent random variables. Write $N_2(W)$ for N_2 to indicate that N_2 depends on the choice of the 2-dimensional Brownian motion W. If θ is an orthogonal transformation on $\mathbf{R}^2$, then $\theta(W)$ is also a 2-dimensional Brownian motion. We check easily that

$$\int_0^{T'} W_t^1 \circ dW_t^2 - \int_0^{T'} W_t^2 \circ dW_t^1$$

is invariant under orthogonal transformation, as is T'. Therefore $N_2(\theta(W)) = N_2(W)$. Since $W_{T'}$ is uniformly distributed on S_r, we can conclude that N_1 and N_2 are independent random variables. This proves the independence property (2).

References

[C] M. CRANSTON. Private notes.

[CH] R. COURANT and D. HILBERT. *Methods of mathematical physics, vol. 2.* Interscience Publishers, New York, 1953.

[D] J. L. DOOB. *Classical potential theory and its probabilistic counterpart.* Grundlehren der mathematischen Wissenshaften **262**, Springer-Verlag, New York, 1984.

[GP] A. GRAY and M. A. PINSKY. The mean exit time from a small geodesic ball in a Riemannian manifold. *Bull. Sci. Math.* (2) **107** (1983), 345-370.

[H] H. R. HUGHES. *Hitting time and place to small geodesic spheres on Riemannian manifolds.* PhD dissertation, Northwestern University, 1988.

[IW] N. IKEDA and S. WATANABE. *Stochastic differential equations and diffusion processes.* North Holland Pub. Co., Amsterdam, New York, Kodansha Ltd., Tokyo, 1981.

[KO] M. KOZAKI and Y. OGURA. On the independence of exit time and exit position from small geodesic balls in Riemannian manifolds. To appear.

[L] M. LIAO. Hitting distributions of small geodesic spheres. *Ann. Prob.* **16** (1988), 1039-1050.

Harry Randolph Hughes
Dept. of Mathematics
Northwestern University
Evanston, Illinois 60201, USA.

Ming Liao
Dept. of Mathematics
Nankai University
Tianjin, P. R. China

THE EXACT HAUSDORFF MEASURE OF BROWNIAN MULTIPLE POINTS, II.

by

Jean-François Le Gall

The purpose of this note is to sharpen a result established in [5] concerning the Hausdorff measure of the set of multiple points of a d-dimensional Brownian motion. Let $X = (X_t, t \geq 0)$ denote a standard two-dimensional Brownian motion and, for every integer $k \geq 1$, let M_k denote the set of k-multiple points of X (a point z is said to be k-multiple if there exist k distinct times $0 \leq t_1 < \ldots < t_k$ such that $X_{t_1} = \ldots = X_{t_k} = z$). A canonical measure on M_k can be constructed as follows. Set:

$$\mathcal{T}_k = \left\{ (t_1, \ldots, t_k) \in (\mathbb{R}_+)^k \ ; \ 0 \leq t_1 < \ldots < t_k \right\}$$

The intersection local time of X with itself, at the order k , is the Radon measure on $\mathcal{T}_k$ formally defined by:

$$\alpha_k(dt_1 \ldots dt_k) = \delta_{(0)}(X_{t_1} - X_{t_2}) \ldots \delta_{(0)}(X_{t_{k-1}} - X_{t_k})\, dt_1 \ldots dt_k$$

where $\delta_{(0)}$ denotes the Dirac measure at 0 in $\mathbb{R}^2$. A precise definition of α_k may be found in Rosen [7] or Dynkin [2]. As the previous formal definition suggests, the measure α_k is supported on the set $\{(t_1, \ldots, t_k); X_{t_1} = \ldots = X_{t_k}\}$ of k-multiple times. Let ℓ_k denote the image measure of α_k by the mapping $(t_1, \ldots, t_k) \longrightarrow X_{t_1}$. It follows that ℓ_k is supported on M_k . Notice that ℓ_k is not a Radon measure, but is a countable sum of finite measures.

The goal of the paper [5] was to obtain upper and lower bounds for the Hausdorff measure (relative to a suitable measure function) of subsets of M_k, in terms of the ℓ_k-measure of these subsets. Set:

$$h_k(x) = x^2 \left(\log \frac{1}{x} \text{ logloglog } \frac{1}{x}\right)^k$$

The main result of [5] gives the existence of two positive constants C_k, C'_k such that, a.s. for every Borel subset F of $\mathbb{R}^2$,

$$(1) \qquad C_k \ell_k(F) \le h_k\text{-}m(F \cap M_k) \le C'_k \ell_k(F)$$

where h_k-m denotes the Hausdorff measure associated with h_k. We improve this result by the following theorem.

THEOREM 1. *There exists a positive constant* c_k *such that, almost surely for every Borel subset* F *of* $\mathbb{R}^2$,

$$\ell_k(F) = c_k \, h_k\text{-}m(F \cap M_k) \; .$$

Theorem 1 shows that the measure ℓ_k is completely determined by the set M_k of k-multiple points. This should be compared with the analogous result of Perkins [6] for the (ordinary) local time at 0 of a linear Brownian motion. When $k = 1$, M_k is the whole Brownian path, and the measure ℓ_k is the Brownian occupation measure. In this particular case, the result of Theorem 1 was obtained by Taylor [8].

PROOF. For simplicity, we assume that $k \ge 2$. We proceed to a first reduction of the problem following [5]. Let I be a subset of $\mathcal{T}_k$ of the type:

$$I = (a_1, b_1) \times \ldots \times (a_k, b_k)$$

where $0 \le a_1 < b_1 < \ldots < a_k < b_k$. The restriction of α_k to I is denoted by α_k^I. The measure α_k^I is finite, and so is ℓ_k^I, defined as the image measure of α_k^I by the mapping $(t_1, \ldots, t_k) \longrightarrow X_{t_1}$. The arguments of [5] show that it is enough to prove the statement of Theorem 1 with ℓ_k replaced by ℓ_k^I and M_k replaced by:

$$M_k^I = \left\{ z \in \mathbb{R}^2;\ z = X_{t_1} = \ldots = X_{t_k} \quad \text{for some} \quad (t_1,\ldots,t_k) \in I \right\}$$

We now apply the following result of geometric measure theory, which may be found in Federer [3, p. 153-155] in a much more general setting. Let μ, ν be two finite Borel measures on $\mathbb{R}^d$. Assume that ν is absolutely continuous with respect to μ. For every $x \in \mathbb{R}^d$, set:

$$(2) \qquad g(x) = \limsup_{r \to 0} \frac{\nu(B(x,r))}{\mu(B(x,r))}$$

where $B(x,r)$ denotes the open ball of radius r centered at x. Then, g is a version of the Radon-Nikodym derivative of ν with respect to μ. We apply this result with $\mu = \ell_k^I$ and ν defined by $\nu(F) = h_k\text{-}m(F \cap M_k^I)$, for any Borel subset F of $\mathbb{R}^2$. The bounds (1) show that μ and ν satisfy the previous assumptions. Therefore, in order to prove Theorem 1, it is enough to check that the function g defined by (2) coincides μ-a.e. with some constant $c_k > 0$. Thus, we must prove that:

$$(4) \qquad \limsup_{r \to 0} \frac{h_k\text{-}m(B(x,r) \cap M_k^I)}{\ell_k^I(B(x,r))} = c_k \quad , \ \ell_k^I(dx)\text{-a.e.} \ , \ P\text{-a.s.}$$

or equivalently,

$$(4') \qquad \limsup_{r \to 0} \frac{h_k\text{-}m(B(X_{t_1},r) \cap M_k^I)}{\alpha_k^I(\{(s_1,\ldots,s_k); X_{s_1} \in B(X_{t_1},r)\})} = c_k \ ,$$

$\alpha_k^I(dt_1..dt_k)$-a.e. , P-a.s. Notice that, by (1) (or more precisely by the analogue of (1) with M_k, ℓ_k replaced by M_k^I, ℓ_k^I, see [5]) the left-hand side of (4') is bounded below and above by positive constants.

The result (4') relates to the behavior of the process X between the successive hitting times of a given k-multiple point. Since we are interested in proving that this result holds for α_k^I-almost all k-multiple times, we may use the results of [4] to replace (4') by an equivalent statement involving intersections of k independent planar Brownian motions defined on the time interval $[-1,1]$. We consider $2k$ independent planar Brownian motions $X'^1,\dots,X'^k$, $X''^1,\dots,X''^k$, starting at 0, and for every $i \in \{1,..,k\}$ we set: $X^i_t = X'^i_t$ if $0 \le t \le 1$, $X^i_t = X''^i_{-t}$ if $-1 \le t < 0$. The intersection local time of the k independent processes $X^1,\dots,X^k$ is the random measure on $[-1,1]^k$ formally defined by:

$$\beta_k(dt_1\dots dt_k) = \delta_{(0)}(X^1_{t_1}-X^2_{t_2}) \dots \delta_{(0)}(X^{k-1}_{t_{k-1}}-X^k_{t_k})\, dt_1\dots dt_k .$$

For every $\rho \in (0,1]$, set:

$$D^\rho_k = \{ z \in \mathbb{R}^2;\ z = X^1_{t_1} = \dots = X^k_{t_k} \text{ for some } t_1,\dots,t_k \in [-\rho,\rho]\} .$$

An easy application of the corollary 2.3 of [4] (see [5] for similar arguments) shows that (4') is equivalent to:

$$\text{(5)} \qquad \limsup_{r\to 0} \frac{h_k\text{-}m(B(0,r) \cap D^\rho_k)}{\beta_k(\{(s_1,\dots,s_k) \in [-\rho,\rho]^k;\ X^1_{s_1} \in B(0,r)\})} = c_k \ , \ \text{P-a.s.}$$

Clearly, the left-hand side of (5) does not depend on the choice of $\rho \in (0,1]$. By the Blumenthal zero-one law applied to the process $(X'^1,\dots,X''^k)$, this quantity is almost surely equal to some constant $c_k \in [0,\infty]$. The remark following (4') shows that $c_k \in (0,\infty)$. This completes the proof of (4') and of Theorem 1. □

Assume now that X is a Brownian motion in $\mathbb{R}^3$ and, for $k = 1$ or 2, let M_k be the set of k-multiple points of X. The intersection local time of X with itself provides a measure on M_k, denoted as above by ℓ_k. Set:

$$h_k(x) = x^{3-k} \left(\log\log \frac{1}{x}\right)^k$$

The next theorem follows from the results of [5] by the same arguments as in the above proof of Theorem 1. Again the case $k = 1$ is due to

Ciesielski and Taylor [1].

THEOREM 2. *For* $k = 1$ *or* 2 *, there exists a positive constant* c_k *such that, almost surely for every Borel subset* F *of* $\mathbb{R}^3$,

$$h_k\text{-}m(F \cap M_k) = c_k\, \ell_k(F).$$

ACKNOWLEDGMENT. It is a pleasure to thank Ed Perkins for pointing out that a suitable density theorem could be used to sharpen our previous results.

REFERENCES.

[1] Ciesielski,Z. Taylor,S.J. : First passage times and sojourn times for Brownian motion in space and the exact Hausdorff measure of the sample path. *Trans. Amer. Math. Soc.* **103**, 434-450 (1962)

[2] Dynkin,E.B. : Random fields associated with multiple points of the Brownian motion. *J. Funct. Anal.* **62**, 397-434 (1985)

[3] Federer,H. : *Geometric Measure Theory* . Springer-Verlag, Berlin Heidelberg New-York 1969.

[4] Le Gall,J.F. : Le comportement du mouvement brownien entre les deux instants où il passe par un point double. *J. Funct. Anal.* **71**, 246-262 (1987)

[5] Le Gall,J.F. : The exact Hausdorff measure of Brownian multiple points. *Seminar on Stochastic Processes 1986* , 107-137. Birkhäuser Boston 1987.

[6] Perkins,E.A. : The exact Hausdorff measure of the level sets of Brownian motion. *Z. Wahrsch. verw. Gebiete* **58**, 373-388 (1981)

[7] Rosen,J. : Self-intersections of random fields. *Ann. Probab.* **12**, 108-119 (1984)

[8] Taylor,S.J. : The exact Hausdorff measure of the sample path for planar Brownian motion. *Proc. Camb. Philos. Soc.* **60**,253-258 (1964)

J.F. Le Gall
Laboratoire de Probabilités de l'Univ. Paris 6
4, Place Jussieu, Tour 56
F-75252 PARIS Cedex 05

On a Stability Property of Harmonic Measures

Peter March*

The purpose of this note is to prove that the exit distributions of a diffusion from a somewhat smooth domain are stable under a large class of perturbations. That this need not be so in general had been observed already by M.V. Keldyš [5], who constructed a Jordan domain D containing the origin, with the property that if $D_n, n \geq 1$, is any sequence of smooth domains such that

$$\overline{D} \subset D_n \subset D_{n-1} \quad \text{and} \quad \overline{D} = \bigcap_n D_n \tag{1}$$

then the classical harmonic measures $h_{D_n}(0, dy), n \geq 1$, do not converge weakly to the harmonic measure $h_D(0, dy)$. Of course, these measures are the exit distributions of Brownian motion starting from the origin.

On the other hand, it is well-known that if $d_n, n \geq 1$, is a sequence of domains such that

$$\bar{d}_n \subset d_{n+1} \subset \bar{d}_{n+1} \subset D \quad \text{and} \quad D = \bigcup_n d_n \tag{2}$$

Then $h_{d_n}(x, dy), n \geq 1$, does converge weakly to $h_D(x, dy)$ for each $x \in D$. The crux of the matter is that whereas exit times of Brownian motion from the inner approximations d_n increase up to the exit time from D, the exit times of the outer approximations D_n decrease down to the exit time of $\overline{D}$ and in general these two times are different. Clearly, wether or not these times are the same is a joint property of the boundary of D and the typical sample path. Thus it develops that the stability of exit distributions of a diffusion from a domain D

* Research supported by an NSERC-Canada grant

under perturbations of domain, generator, etc., will be determined by the almost sure equality of the exit time from D and the slightly larger exit time from $\overline{D}$.

This idea has been emphasized by Stroock [9], in a study of the measurability of entrance times, and by Stroock and Varadhan [10] §5-9 esp. §10, in a stochastic formulation and resolution of the first boundary value problem for degenerate elliptic equations. And although Keldyš [5] does not take a probabilistic point of view, his results on the stability of the Dirichlet problem for the Laplace operator have a probablistic interpretation as follows:

(3) The exit distributions of Brownian motion from a domain D are stable under perturbations of D if and only if for each $x \in D$ the exit times from D and $\overline{D}$ are P_x-almost surely equal.

Let us gather together those notations and definitions needed to state the main result. Following the proof is a discussion of (3) and certain generalizations of it.

For our purposes, a domain D is non empty, bounded, open subset of $\mathbf{R}^d$ and a diffusion operator L is a differential operator of the form

$$Lu(x) = \sum_{i,j=1}^{d} a^{ij}(x)\frac{\partial^2}{\partial x_i \partial x_j}u(x) + \sum_{i=1}^{d} b^i(x)\frac{\partial}{\partial x_i}u(x). \tag{4}$$

whose coefficients are bounded continuous functions satisfying the ellipticity condition $\sum_{i,j=1}^{d} a^{ij}(x)\xi_i\xi_j > 0$ for all vectors $x, \xi \in \mathbf{R}^d$. Introduce the space of data

$$\mathcal{D} = \{(D, x, L) : D \text{ a domain, } x \in \overline{D}, L \text{ a diffusion operator}\} \tag{5}$$

We agree to identify two points (D, x, L) and (D, x, L') if there is an open set V containing $\overline{D}$ such that $Lf = L'f$ on V for all test functions $f \in C_c^\infty(\mathbf{R}^d)$. Concerning convergence in $\mathcal{D}$ let us say

(6) A sequence D_n converges to D if for every pair of open sets U, V such that $\overline{U} \subset D \subset \overline{D} \subset V$, the inclusion $\overline{U} \subset D_n \subset \overline{D}_n \subset V$ holds for all sufficiently large n;

(7) A sequence x_n converges to x if $\lim_{u\to\infty} x_n = x$ in $\mathbf{R}^d$;

(8) A sequence L_n converges moderately to L on $\overline{D}$ if there is a bounded open set V containing $\overline{D}$ such that

(a) there is an $\epsilon > 0$ such that $\sum_{i,j=1}^{d} a_n^{ij}(x)\xi_i\xi_j \geq \epsilon|\xi|^2$ for all $x \in V$, $\xi \in \mathbf{R}^d$ and $n \geq 1$;

(b) for $1 \leq i,j \leq d$ the sequences $a_n^{ij}, n \geq 1$, and $b_n^i, n \geq 1$, are uniformly bounded and equicontinuous on $\overline{V}$;

(c) if a^{ij}, b^i are any limit points from (b) above, and L' is the diffusion operator with these coefficients, then $L'f(x) = Lf(x)$ on $\overline{D}$ for all $f \in C_c^\infty(V)$.

Let us say that $(D_n, x_n, L_n), n \geq 1$, converges moderately in $\mathcal{D}$ to (D, x, L) if conditions (6) - (8) hold; and we write $(\mathcal{D}, m)$ for the space $\mathcal{D}$ with this notion of convergence. We discuss the appropriateness of this kind of convergence after the proof of our result.

According to Stroock and Varadhan [11], there is a unique, Feller continuous family of diffusion laws $P_x^L, x \in \mathbf{R}^d$, which is the solution of the martingale problem for L. For any Borel set $A \subset \mathbf{R}^d$, the entrance time

$$\tau_A = \inf\{t > 0 : \ X_t \in A\} \tag{9}$$

is a P_x^L-measurable function and we define the L-harmonic measure of a domain D relative to a point $x \in \overline{D}$ as

$$h_{D,L}(x, dy) = P_x^L\{X(\tau_{D^c}) \in dy; \ \tau_{D^c} < \infty\}. \tag{10}$$

It is known, and we note it below, that if $x \in \overline{D}$ then $P_x^L\{\tau_{D^c} < \infty\} = 1$. Thus we may consider harmonic measure as a map

$$h : (\mathcal{D}, m) \to \mathcal{M}_1, \quad h[(D, x, L)](dy) = h_{D,L}(x, dy) \tag{11}$$

from $\mathcal{D}$ to the space $\mathcal{M}_1$ of probability measures on $\mathbf{R}^d$ with its topology of weak convergence. The continuity of h, as it turns out, is determined by the relationship between the random times

$$e_D = \tau_{D^c} \quad \text{and} \quad \pi_D = \tau_{(\overline{D})^c}; \tag{12}$$

the so-called exit and penetration times respectively. Our main result is the following

Theorem. The function $h : (\mathcal{D}, m) \to \mathcal{M}_1$ is continuous at a point (D, x, L) if D is a Lipshitz domain.

After the proof we discuss several aspects of this theorem including two generalizations of the Keldyš result (3). However, before proceeding we want to emphasize that our proofs have been simplified by liberal use of the ideas and exercises of Stroock and Varadhan [11].

We begin by stating several known results which are important for the proof of the theorem; proofs or references are included for the reader's convenience. In what follows, (D, x, L) is a typical point of $\mathcal{D}$.

Proposition 1.

(a) π_D is an upper semicontinuous function of $\omega \in C([0,\infty), \mathbf{R}^d)$, relative to the topology of uniform convergence on bounded time intervals;

(b) e_D is a lower semicontinuous function of ω if and only if either $\omega(0) \notin \partial D$ or both $\omega(0) \in \partial D$ and $e_D(\omega) = 0$;

(c) for each ω, $\quad 0 \le e_D \le \pi_D$.

Proof: (a) If $\pi_D(\omega) = +\infty$ then there is nothing to prove; so suppose $\pi_D(\omega) < T < +\infty$ and let $\omega_n, n \ge 1$ converge to ω uniformly on $[0, T]$. There is a time s, with $\pi_D(\omega) < s < T$ such that $\omega(s) \in (\overline{D})^c$ and since this set is open, $\mathrm{dist}(\omega(s), \overline{D}) = \delta > 0$. Then for all suitably large n, $|\omega_n(s) - \omega(s)| < \frac{1}{2}\delta$, hence $\pi_D(\omega_n) \le s < T$. Since $T > \pi_D(\omega)$ was arbitrary, this implies, $\limsup_{n\to\infty} \pi_D(\omega_n) \le \pi_D(\omega)$, which was to be proved.

(b) If $\omega(0) \notin \partial D$ the proof is entirely similar to item (a), and if $\omega(0) \in \partial D$ and $e_D(\omega) = 0$ the proof is trivial. If $\omega(0) \in \partial D$ and $e_D(\omega) > 0$ then there is a sequence of continuous functions $\lambda_n, n \ge 1$ of the form

$$\lambda_n(t) = \begin{cases} 0, & \text{if } 0 \le t \le \epsilon_n \\ t, & \text{if } t \ge 2\epsilon_n \end{cases}$$

such that for suitably chosen $\epsilon_n \downarrow 0$, the sequence $\omega_n = \omega \circ \lambda_n$ converges uniformly to ω. Yet $e_D(\omega_n) = 0$ while $e_D(\omega) > 0$, showing that, in this case, e_D is not lower semi continuous.

The proof of (c) is obvious, and so omitted. □

Proposition 2. Let D be a domain.

a) There exist sequences of domains $d_k, k \geq 1$ and $\delta_k, k \geq 1$ such that $\overline{D} \subset \delta_{k+1} \subset \bar{\delta}_{k+1} \subset \delta_k$ and $\overline{D} = \bigcap_k \delta_k$ and, as well, $\bar{d}_k \subset d_{k+1} \subset \bar{d}_{k+1} \subset D$ and $D = \bigcup_k d_k$.

b) Let $D_n, n \geq 1$ be a sequence of domains converging to D as in (6). Then (i) $\limsup_{n\to\infty} \pi_{D_n} \leq \pi_D$ and (ii) $e_D \leq \liminf_{n\to\infty} e_{D_n}$ if and only if either $\omega(0) \notin \partial D$ or both $\omega(0) \in \partial D$ and $e_D(\omega) = 0$.

Proof: The proof of item (a) is elementary and so, omitted. As for (b), let's suppose $\omega(0) \in D$, and let $0 < T < e_D(\omega)$. Then the trace $X_{[0,T]} = \{X(t) : 0 \leq t \leq T\}$ is a compact subset of D, and so there is an open subset U of D such that $X_{[0,T]} \subset U \subset \overline{U} \subset D$. Thus $X_{[0,T]} \subset D_n$ for all sufficiently large n, say $n \geq m$, and so $T \leq \inf_{n\geq m} e_{D_n}$. Since $T < e_D$ was arbitrary, it follows that $e_D \leq \liminf_{n\to\infty} e_{D_n}$. The remaining cases and the statement concerning π_D are proved in entirely similar ways and so their proofs are omitted. □

Proposition 3. Let U be a domain, and let L_n a sequence of diffusion operators with coefficients a_n^{ij} and b_n^i. Let $\lambda > 0$ and $T > 0$ be given and define

$$\gamma_{\lambda,T} = \sup\{|X_t - X_s| : \quad 0 \leq s \leq t \leq T \text{ and } t - s < \lambda\}.$$

Suppose there are constants k_1 and k_2 such that for all $1 \leq i, j \leq d$, $n \geq 1$, and $x, \xi \in \mathbf{R}^d$,

$$|b_n^i(x)| \leq k_1 \quad \text{and} \quad k_2^{-1}|\xi|^2 \leq a_n^{ij}(x)\xi_i\xi_j \leq k_2|\xi|^2.$$

Then

$$(a) \qquad \sup_n \sup_{x\in\overline{U}} P_x^{L_n}\{e_U\} < \infty; \quad \text{and}$$

$$(b) \qquad \lim_{\lambda\to 0} \sup_n \sup_{x\in\overline{U}} P_x^{L_n}\{\gamma_{\lambda,T} > \rho\} = 0.$$

Proof: These are two basic integrability results in the martingale approach to diffusion theory. See Stroock and Varadhan [11] p 251 exercise 10.3.1 for item (a) and [11] pp. 36-40 for item (b). □

Proposition 4. Let $(D, x_0, L) \in \mathcal{D}$, let $x_0 \in \partial D$ and suppose there is an open cone Γ with vertex x_0 such that $\overline{\Gamma} \cap \overline{D} = \{x_0\}$. Then $P^L_{x_0}\{\pi_D = 0\} = 1$.

Proof: This proposition can be made to follow from the existence, under the hypothesis above, of a barrier function [7]. Instead, we use a pretty scaling argument employed by Stroock and Varadhan [10] and [11] p.283 exercise 11.5.3.

Define $S_\epsilon : C([0,\infty), \mathbf{R}^d) \to C([0,\infty), \mathbf{R}^d)$ by $(S_\epsilon \omega)(t) = \epsilon^{-1/2}(\omega(\epsilon t) - \omega(0)) + \omega(0)$ and check that the measure $P^L_{x_0} \circ S_\epsilon^{-1}$ solves the martingale problem for the diffusion operator $L^{x_0}_\epsilon$:

$$(13) \qquad L^{x_0}_\epsilon f(x) = \sum_{i,j=1}^d a^{ij}_{\epsilon,x_0}(x) \frac{\partial^2}{\partial x_i \partial x_j} f(x) + \sum_{i=1}^d b^i_{\epsilon,x_0}(x) \frac{\partial}{\partial x_i} f(x)$$

where

$$a^{ij}_{\epsilon,x_0}(x) = a^{ij}(x_0 + \sqrt{\epsilon}(x - x_0))$$

and

$$b^i_{\epsilon,x_0}(x) = \sqrt{\epsilon} b^i(x_0 + \sqrt{\epsilon}(x - x_0)).$$

Indeed, if $g^{x_0}_\epsilon(x) = \epsilon^{-1/2}(x - x_0) + x_0$, and $f \in C^\infty_c(\mathbf{R}^d)$, then

$$(14) \qquad \begin{aligned} & f \circ g^{x_0}_\epsilon(X_{\epsilon t}) - \int_0^{\epsilon t} L(f \circ g^{x_0}_\epsilon)(X_s) ds \\ =& f \circ g^{x_0}_\epsilon(X_{\epsilon t}) - \int_0^t \epsilon L(f \circ g^{x_0}_\epsilon)(X_{\epsilon s}) ds \\ =& f \circ g^{x_0}_\epsilon(X_{\epsilon t}) - \int_0^t (L^{x_0}_\epsilon f) \circ g^{x_0}_\epsilon(X_{\epsilon s}) ds \end{aligned}$$

is a $P^L_{x_0}$-martingale; so that $P^L_{x_0} \circ S_\epsilon^{-1} = P^{L^{x_0}_\epsilon}_{x_0}$ by the uniqueness of solutions of the martingale problem. By [11] Theorem 1.4.6, the family $\left\{ P^{L^{x_0}_\epsilon}_{x_0} ; \epsilon > 0 \right\}$ is tight, and also $L^{x_0}_\epsilon f$ converges to $L^{x_0}_0 f$ for all $f \in C^\infty_c(\mathbf{R}^d)$. Thus $P^L_{x_0} \circ S_\epsilon^{-1}$ converges weakly to $P^{L^{x_0}_0}_{x_0}$.

Because $\Gamma - x_0$ is invariant under dilations we have $P^L_{x_0}\{X_\epsilon \in \Gamma\} = P^L_{x_0} \circ S_\epsilon^{-1}\{X_1 \in \Gamma\}$; and because Γ is open, so is the set of paths $\{\omega : \omega(1) \in \Gamma\}$. Thus

$$(15) \qquad \begin{aligned} \liminf_{\epsilon \to 0} P^L_{x_0}\{X_\epsilon \in \Gamma\} &= \liminf_{\epsilon \to 0} P^L_{x_0} \circ S_\epsilon^{-1}\{X_1 \in \Gamma\} \\ &\geq P^{L^{x_0}_0}_{x_0}\{X_1 \in \Gamma\} > 0. \end{aligned}$$

Now, $\{\pi_D = 0\} = \bigcap_{\epsilon > 0} \{\pi_D < 2\epsilon\}$ and if $x_0 \in \partial D$ then

$$(16) \qquad P^L_{x_0}\{\pi_D < 2\epsilon\} \geq P^L_{x_0}\{\tau_\Gamma < 2\epsilon\} \geq P^L_{x_0}\{X_\epsilon \in \Gamma\}$$

Thus by (15),

(17) $$P^L_{x_0}\{\pi_D = 0\} \geq \liminf_{\epsilon\to 0} P^L_{x_0}\{X_\epsilon \in \Gamma\} > 0,$$

so that, in fact, $P^L_{x_0}\{\pi_D = 0\} = 1$ according to Blumenthal's zero-one law. □

We come now to the proof of the main result.

Theorem 1. The function $h : (\mathcal{D}, m) \to \mathcal{M}_1$ is continuous at a point (D, x, L) if D is a Lipshitz domain.

Proof: Let D be a Lipshitz domain and $x_0 \in \partial D$. There is an open truncated cone with vertex x_0 contained in $(\overline{D})^c$ and, as our considerations are of a local nature, it is no loss of generality to suppose this cone is not in fact truncated but rather a full cone as in Proposition 4. If L' is a diffusion operator then by Proposition 4 and the Markov property

(18) $$P^{L'}_x\{e_D = \pi_D\} = P^{L'}_x P^{L'}_{X(e_D)}\{\pi_D = 0\} = 1.$$

for all $x \in \overline{D}$. The proof of the theorem depends on this fact and its consequence, that for all $x \in \overline{D}$, the exit time e_D is a $P^{L'}_x$-almost surely continuous function of ω.

Let ϕ be a bounded continuous function on $\mathbf{R}^d$ and suppose (D_n, x_n, L_n), $n \geq 1$, converges in $(\mathcal{D}, m)$ to (D, x, L). Set

(19) $$\alpha = \liminf_{n\to\infty} \int \phi(y) h_{D_n,L_n}(x_n, dy) \quad \text{and}$$
$$\beta = \limsup_{n\to\infty} \int \phi(y) h_{D_n,L_n}(x_n, dy)$$

and let us show that

(20) $$\alpha = \beta = \int \phi(y) h_{D,L}(x, dy).$$

Coming first to α, let n' be a subsequence along which the limit inferior is achieved. By (8), there is a further subsequence, which we also denote n', along which the coefficients converge, namely

(21) $$\tilde{a}^{ij} = \lim_{n'\to\infty} a^{ij}_{n'}, \quad \tilde{b}^i = \lim_{n'\to\infty} b^i_{n'}$$

uniformly on a bounded open set V containing $\overline{D}$. Also by (8), $\tilde{a}^{ij}$ and $\tilde{b}^{ij}$ are bounded continuous functions satisfying the ellipticity condition in line 8(a). Let us assume that the $\tilde{a}^{ij}, \tilde{b}^{ij}$ have been extended to bounded, continuous functions on $\mathbf{R}^d$ with

(22) $$\sum_{i,j=1}^{d} \tilde{a}^{ij}(x)\xi_i\xi_j \geq \epsilon|\xi|^2 \quad \text{for all } x \in \mathbf{R}^d.$$

If $\tilde{L}$ is the diffusion operator with these coefficients then $L_{n'}f$ converges uniformly to $\tilde{L}f$ for all $f \in C_c^\infty(V)$. Since, by hypothesis, $\tilde{L}f = Lf$ on $\overline{D}$ it follows, [11] exercise 11.5.1, p. 283, that

(23) $$P_x^{\tilde{L}}|_{\mathcal{F}_{e_D}} = P_x^{L}|_{\mathcal{F}_{e_D}}$$

for each $x \in D$; in particular

(24) $$h_{D,\tilde{L}}(x, dy) = h_{D,L}(x, dy).$$

Of course $h_{D,\tilde{L}}(x, dy) = h_{D,L}(x, dy) = \delta_x(dy)$ if $x \in \partial D$ by Proposition 4, so that the harmonic measures agree in this case too. Thus it is clear we may, and so do, assume that $L_{n'}f$ converges uniformly to $\tilde{L}f$ for all $f \in C_c^\infty(\mathbf{R}^d)$, that the coefficients of $L_{n'}$ and $\tilde{L}$ are uniformly bounded and satisfy (22). Thus, [11], the diffusion laws $\left\{P_y^{L_{n'}}; n \geq 1, \quad y \in K\right\}$, K any compact set, are tight and $\lim_{n' \to \infty} P_y^{L_{n'}} = P_y^{\tilde{L}}$ and $\lim_{n' \to \infty} P_{x_{n'}}^{L_{n'}} = P_x^{\tilde{L}}$.

For notational convenience let us agree to drop the prime superscript from the subseqence n'. We have then

(25) $$\int \phi(y) h_{D_n, L_n}(x_n, dy) = P_{x_n}^{L_n}\{\phi(X_{e_{D_n}})\}$$

$$= P_{x_n}^{L_n}\{\phi(X_{e_D})\} + P_{x_n}^{L_n}\{\phi(X_{e_{D_n}}) - \phi(X_{e_D})\}$$

Since $\omega \to \phi(X_{e_D})$ is $P_x^{\tilde{L}}$-almost surely continuous, it follows from (24) that

(26) $$\lim_{n \to \infty} P_{x_n}^{L_n}\{\phi(X_{e_D})\} = P_x^{\tilde{L}}\{\phi(X_{e_D})\}$$
$$= \int \phi(y) h_{D,L}(x, dy)$$

Let $M_n = \left|\phi(Xe_{D_n}) - \phi(X_{e_D})\right|$ and let

(27) $$\eta_\phi^V(\rho) = \sup\left\{|\phi(x) - \phi(y)| : x, y \in \overline{V}, \quad |x - y| \leq \rho\right\}$$

be a uniform modulus of continuity for ϕ. It remains to show that $\limsup_{n\to\infty} P_{x_n}^{L_n}\{M_n\} = 0$. To this end let $T > 0, \lambda > 0, \rho > 0$ be given and let $d_k, k \geq 1$, and $\delta_k, k \geq 1$, be domains such that

$$\bar{d}_k \subset d_{k+1} \subset \bar{d}_{k+1} \subset D, \quad D = \bigcup_{k\geq 1} d_k \quad \text{and} \tag{28}$$

$$\overline{D} \subset \delta_k \subset \bar{\delta}_k \subset \delta_{k-1} \subset V, \quad \overline{D} = \bigcap_{k\geq 1} \delta_k.$$

Then for fixed k and all suitably large n,

$$\begin{aligned} P_{x_n}^{L_n}\{M_n\} &= P_{x_n}^{L_n}\{M_n; e_V \leq T, \pi_{\delta_k} - e_{d_k} < \lambda, \gamma_{\lambda,T} \leq \rho\} \\ &\quad + P_{x_n}^{L_n}\{M_n; e_V \leq T, \pi_{\delta_k} - e_{d_k} < \lambda, \gamma_{\lambda,T} > \rho\} \\ &\quad + P_{x_n}^{L_n}\{M_n; e_V \leq T, \ \pi_{\delta_k} - e_{d_k} \geq \lambda\} \\ &\quad + P_{x_n}^{L_n}\{M_n; e_V > T\} \\ &\leq \eta_\phi^V(\rho) + 2\|\phi\|_\infty P_{x_n}^{L_n}\{\pi_{\delta_k} - e_{d_k} \geq \lambda\} \\ &\quad + 2\|\phi\|_\infty \left[\sup_n P_{x_n}^{L_n}\{\gamma_{\lambda,T} > \rho\} + \sup_n P_{x_n}^{L_n}\{e_V > T\}\right]. \end{aligned} \tag{29}$$

Since the domains d_k are chosen by us, we may assume that for each fixed $k \geq 1$ there is a $\theta > 0$ such that dist $(x_n, \partial d_k) \geq \theta$ for $n \geq 1$. If $\Lambda \equiv \{\omega : \text{dist}\,(\omega(0), \partial d_k) \geq \frac{1}{2}\theta\} \subset C\left([0,\infty), \mathbf{R}^d\right)$ then by Proposition 1, $\{\pi_{\delta_k} - e_{d_k} \geq \lambda\} \cap \Lambda$ is a closed subset of Λ. Since $P_{x_n}^{L_n} = P_{x_n}^{L_n}\big|_\Lambda$ it follows that

$$\limsup_{n\to\infty} P_{x_n}^{L_n}\{\pi_{\delta_k} - e_{d_k} \geq \lambda\} \leq P_x^{\bar{L}}\{\pi_{\delta_k} - e_{d_k} \geq \lambda\}. \tag{30}$$

By Proposition 2, and line (18),

$$\lim_{k\to\infty} P_x^{\bar{L}}\{\pi_{\delta_k} - e_{d_k} \geq \lambda\} = P_x^{\bar{L}}\{\pi_D - e_D \geq \lambda\} = 0. \tag{31}$$

Thus by (29) - (31),

$$\limsup_{n\to\infty} P_{x_n}^{L_n}\{M_n\} \leq \eta_\phi^V(\rho) + 2\|\phi\|_\infty \left[\sup_n P_{x_n}^{L_n}\{\gamma_{\lambda,T} > \rho\} + \sup_n P_{x_n}^{L_n}\{e_V > T\}\right].$$

Now let $\lambda \to 0$ then let $T \to \infty$ and $\rho \to 0$, which yields $\limsup_{n\to\infty} P_{x_n}^{L_n}\{M_n\} = 0$ by Proposition 3.

Thus $\alpha \equiv \liminf_{n\to\infty} \int \phi(y) h_{D_n,L_n}(x_n, dy) = \int \phi(y) h_{D,L}(x, dy)$. Identical reasoning shows that $\beta = \int \phi(y) d_{D,L}(x, dy)$ as well; and so the theorem is proved.

□

It should be pointed out again that the theorem hinges on the fact that if D is a Lipshitz domain then $P_x^L\{e_D = \pi_D\} = 1$ for $x \in D$ and $P_x^L\{\pi_D = 0\} = 1$ for $x \in \partial D$ for any diffusion operator L. We say that a point $x \in \partial D$ is L-regular if $P_x^L\{e_D = 0\} = 1$ and, following Keldyš, that x is L-stable if $P_x^L\{\pi_D = 0\} = 1$. P. Bauman [1] has given a Wiener-type criterion for L-regularity and as a consequence for L-stability of a boundary point. Examples show that different diffusion operators may not agree on the regularity or stability of a given boundary point [8], see also [6]; however, there is agreement among those operators whose second order coefficients are Hölder continuous [3]. (This also follows, for example from the comparibility of Green functions [4].) Thus we may strengthen Keldyš statement (3), see [5] Theorem XVI, as follows. Let $\mathcal{D}_\alpha \subset \mathcal{D}$ be the set of points (D, x, L) such that the second order coefficients of L are α-Hölder continuous. We say a sequence $(D_n, x_n, L_n), n \geq 1$, converges moderately to (D, x, L) in $\mathcal{D}_\alpha$ if it converges moderately to (D, x, L) in $\mathcal{D}$ and the following condition holds:

(32) There is a bounded open set V containing $\overline{D}$ and a $K > 0$ such that $|a_n^{ij}(x) - a_n^{ij}(y)| \leq K|x - y|^\alpha$ for all $1 \leq i, j \leq d$ and $x, y \in V$.

We write $(\mathcal{D}_\alpha, m)$ for $\mathcal{D}_\alpha$ with this notion of convergence.

Theorem 2 Let $(D, x, L) \in \mathcal{D}_\alpha$ and let $s(D)$ be the set of boundary points of D which are stable for the Laplace operator.

a) Suppose $x \in \partial D$. Then $h : (\mathcal{D}_\alpha, m) \to \mathcal{M}_1$ is continuous at (D, x, L) if and only if $x \in s(D)$.

b) The function $h : (\mathcal{D}_\alpha, m) \to \mathcal{M}_1$ is continuous at (D, x, L) for every $x \in D$ if and only if $h_{D,L}(x, s(D)) = 1$ for every $x \in D$.

The sufficiency of these conditions is exactly as in the proof of Theorem 1, using in place of Proposition 4, the fact that in the Hölder continuous case, a point x is L-stable if and only if $x \in s(D)$. To see the necessity in item (a) note that if h is continuous at (D, x, L) then, as usual, x must be a regular point. (Take $D_n = D, L_n = L, x_n \in D, x_n \to x$). In this case let $L_n = L, x_n = x$ and $D_n, n \geq 1$ a sequence of domains such that

$$\overline{D} \subset D_n \subset \overline{D}_n \subset D_{n-1} \quad \text{and} \quad \overline{D} = \bigcap_n D_n. \tag{33}$$

Proposition 2 and the continuity of h gives the identity

$$\phi(x) = P_x^L\{\phi(X_{e_D})\} = P_x^L\{\phi(X_{\pi_D})\} \tag{34}$$

for all bounded continuous ϕ and it follows by standard reasoning that $P_x^L\{\pi_D = 0\} = 1$; that is $x \in s(D)$. Part (b) follows readily from the similar identity

$$P_x^L\{\phi(X_{e_D})\} = P_x^L\{\phi(X_{\pi_D})\} = P_x^L P_{X(e_D)}^L\{\phi(X_{\pi_D})\} \tag{35}$$

for all $x \in D$ and bounded continuous ϕ.

If we strengthen the notion of convergence of diffusion operators we can obtain a stability statement with no further restrictions on the domain D or the operator L. Let us say that $(D_n, x_n, L_n), n \geq 1$, converges strongly to (D, x, L) in $\mathcal{D}$ if (6) and (7) hold and if for all $f \in C_c^\infty(\mathbf{R}^d)$ $L_n f$ converges uniformly to Lf on an open set V containing $\overline{D}$. We write $(\mathcal{D}, s)$ for $\mathcal{D}$ with this notion of convergence. The following result is not hard to show, given the preceeding arguments.

Theorem 3. Let $(D, x, L) \in \mathcal{D}$.

(a) Suppose $x \in \partial D$. Then $h : (\mathcal{D}, s) \to \mathcal{M}_1$ is continuous at (D, x, L) if and only if $P_x^L\{\pi_D = 0\} = 1$.

(b) The function $h : (\mathcal{D}, s) \to \mathcal{M}_1$ is continuous at (D, x, L) for all $x \in D$ if and only if $P_x^L\{e_D = \pi_D\} = 1$ for all $x \in D$.

To try to check these conditions in a particular case one may refer to Bauman's Wiener-type criterion [1]. However there are trivial cases where they fail to hold; for example if D is the slit disk in the plane, no point on the open slit is an L-stable boundary point for any L. (It is interesting to note, however, that the exit time e_D is still P_x^L-almost surely continuous even though if $x \in D$ then $P_x^L\{e_D = \pi_D\} < 1$.)

We close this note by sketching an example which shows that if it is only assumed that $L_n f$ converges to Lf uniformly on $\overline{D}$ then h need not be continuous at $(D, \cdot, L)$. Thus some conditions of the form (8) are needed for the stability of harmonic measures. On the other hand, Caratheodory's notion of the kernel of a sequence of domains and of convergence of the sequence to its kernel, which is appropriate for the stability of conformal maps, [2] pp. 74-77, is a weaker type of convergence than that imposed in (6).

Our example is the following. Let

$$D = B(0,1) \subset \mathbf{R}^d\,, \quad D_n = B(0,1+1/n), \quad Q = (0,\ldots,0,1) \tag{36}$$
$$Q_n = (0,\ldots,0,1+1/n)\,, \quad \text{and} \quad L + 1/2\Delta.$$

The operators L_n are smooth modifications of the generator of Brownian motion conditioned to exit D_n at Q_n.

Let $1 < r_n < \rho_n < 1 + 1/n$ be constants to be chosen conveniently later on. Let $\psi_n, \phi_n : \mathbf{R}_+ \to [0,1]$ be smooth cut off functions such that

$$\psi_n(r) = \begin{cases} 0, & \text{if } 0 \le r \le \frac{1}{2}(1+r_n) \\ 1, & \text{if } r \ge r_n \end{cases} \tag{37}$$

and

$$\phi_n(r) = \begin{cases} 1, & \text{if } 0 \le r \le \rho_n \\ 0, & \text{if } r \ge \frac{1}{2}(1+1/n+\rho_n) \end{cases} \tag{38}$$

Let h_n be the Poisson kernel of $L = \frac{1}{2}\Delta$ in D_n with pole at Q_n. Of course h_n is harmonic and

$$L_n^1 = \frac{1}{2}\Delta + \frac{\nabla h_n(x)}{h_n(x)} \cdot \nabla \tag{39}$$

is the generator of Brownian motion conditioned to exit D_n at Q_n. Let

$$L_n^2 = \frac{1}{2}\Delta + \psi_n(|x|)\frac{\nabla h_n(x)}{h_n(x)} \cdot \nabla \tag{40}$$

and

$$L_n = \frac{1}{2}\Delta + \phi_n(|x|)\psi_n(|x|)\frac{\nabla h_n(x)}{h_n(x)} \cdot \nabla \tag{41}$$

Now for each $x \in D_n$, $P_x^{L_n^1}\{X_{\zeta}- = Q_n\} = 1$ where ζ is the lifetime of process X. It follows that $P_x^{L_n^2}\{X_{\zeta}- = Q_n\} = 1$ for each $x \in D_n$ because, as the coefficients of L_n^1 and L_n^2 agree on a neighborhood of ∂D_n, the associated diffusions share the same singular final behavior.

Now Q_n is a regular boundary point for L_n and so there is a $\eta_n > 0$ such that

$$\inf_{y \in D_n \cap B(Q_n,\eta_n)} P_x^{L_n}\{X(e_{D_n}) \in B(Q_n, \frac{1}{n})\} \ge 1 - 1/n \tag{42}$$

Choose $1 + 1/n - \eta_n < \rho_n < 1 + 1/n$ with ρ_n so close to $1 + 1/n$ that, with $T_n = e_{B(0,\rho_n)}$,

$$P_x^{L_n}\{X_{T_n} \in B(Q_n,\eta_n)\} \equiv P_x^{L_n^2}\{X_{T_n} \in B(Q_n,\eta_n)\} \ge 1 - \frac{1}{n}. \tag{43}$$

uniformly in $x \in \overline{D}$.

Then we have for all $x \in \overline{D}$,

$$
\begin{aligned}
(44) \qquad & P_x^{L_n}\{X(e_{D_n}) \in B(Q_n, \frac{1}{n})\} \\
& \geq P_x^{L_n}\{P_{X_{T_n}}^{L_n}\{X(e_{D_n}) \in B(Q_n, \frac{1}{n})\}; X_{T_n} \in B(Q_n, \eta_n)\} \\
& \geq \left(1 - \frac{1}{n}\right)^2
\end{aligned}
$$

Thus we have D_n converging to $D, L_n f(x) \equiv Lf(x)$ on $\{|x| < \frac{1}{2}(1 + r_n)\} \cup \{|x| > 1/2\,(1 + 1/n + \rho_n)\}$ for every $f \in C_c^\infty(\mathbb{R}^d)$ and as well every $x \in \partial D$ is a stable boundary point. Yet on the strength of (44),

$$
(45) \qquad \lim_{n\to\infty} h_{D_n, L_n}(x, dy) = \delta_Q(dy)
$$

for each $x \in \overline{D}$ whereas $h_{D,L}(x, dy)$ is the classical harmonic measure of the unit ball which, in particular, has a density relative to Lebesgue measure on ∂D if $x \in D$.

References

[1.] P. Bauman, A wiener test for nondivergence structure, second order elliptic equations, Indiana U. Math. J., 34, (1985), 825-844.

[2.] C. Caratheodory, Conformal Representation, Cambridge Tracts in Mathematics and Mathematical Physics, Cambridge University Press, London, second edition, reprinted, 1958.

[3.] R.M. Hervé, Recherches axiomatiques sur la theorie des fonctions surharmoniques et du potential, Ann. Inst. Fourier (Grenoble), 12, (1962), 415-571.

[4.] H. Hueber and M. Sieveking, Uniform bounds for quotients of green functions on $C^{1,1}$−domains, Ann. Inst. Fourier, 32, (1982), 105-117.

[5.] M.V. Keldyš, On the solvability and stability of the dirichlet problem, A.M.S. Translations, Series 2, Vol. 51, 1966.

[6.] E.M. Landis, s−capacity and the behavior of a solution of a second order elliptic equation with discontinuous coefficients in the neighbourhood of a boundary point, Soviet Math. Dokl., 9 (1968), 582-586.

[7.] K. Miller, Barriers on cones for uniformly elliptic operators, Ann. Math. Pura. Applicata IV (1967), 93-105.

[8.] ——, Non equivalence of regular boundary points for the Laplace and non divergence equations, even with continuous coefficients, Ann. Scuola. Norm. Sup. Pisa, (3), 24, 1970, 159- 163.

[9.] D.W. Stroock, Penetration times and passage times, in Markov Processes and Potential Theory, J. Chover, ed., Publ. 19, M.R.C., U. of Wisconsin, J. Wiley and Son, 1967.

[10]. D.W. Stroock and S.R.S. Varadhan, On degenerate elliptic-parabolic operators of second order and their associated diffusions, Comm. Pure Appl. Math., 24, 1972, 651-714.

[11.] ——, Multidimensional Diffusion Processes, Springer-Verlag, Berlin, Heidelberg, New York, 1979.

Peter March

Department of Mathematics & Statistics

Carleton University

Ottawa, CANADA, K1S 5B6

BEHAVIOUR OF EXCESSIVE FUNCTIONS OF CERTAIN DIFFUSIONS UNDER THE ACTION OF THE TRANSITION SEMI-GROUP

by

Z.R. Pop-Stojanović

Introduction. In earlier papers [4],[5], it has been shown that under certain analytic conditions concerning its potential kernel, a strong Markov process, which is transient and with continuous sample paths, has all of its excessive harmonic functions, which are not identically infinite, continuous . Also, it has been shown that under the same conditions the excessiveness of harmonic functions of the process is automatic . In this paper we are studying the behaviour of excessive functions of the process under the action of the transition semi-group of the process. For example, all excessive functions for the Brownian motion semi-group are transformed into continuous functions by the semi-group. It seems that even this classical case does not appear in the literature. This will be shown below under a more general setting.

Setting. In this paper $X = (\Omega, F, F_t, X_t, \theta_t, P^x)$ will denote a transient diffusion, i.e., a strong Markov process with continuous sample paths on a locally compact Hausdorff state space $(E, \mathcal{E})$ with a countable base. Following [2],[3], we are assuming the existence of a potential kernel with the following properties.

Let $U(x,dy) = u(x,y)\xi(dy)$ denote this kernel where ξ is a Radon measure, (which is going to be denoted by dy in the sequel), and the potential density function u is such that:

(a) For every x, and for every y, function $(x,y) \to u^{-1}(x,y)$ is finite and continuous; in particular, this implies $u(x,y) > 0$ for all (x,y).

(b) $u(x,y) = \infty$ if and only if $x = y$.

Recall that

(c) For every nonnegative neasurable function f ,

$$E^x[\int_0^\infty f(X_t)\, dt\,] = \int u(x,y)f(y)\xi(dy)\ .$$

Other notations used here: if (P_t) denotes the transition semi-group of the process X, then

$$P_t u(x,y) = \int P_t(x,dz)u(z,y)\ ;$$

$$P_A u(x,y) = \int P_A(x,dz)u(z,y)\ ,$$

where A is a Borel set in E;

$$P_A f(x) = E^x[f(X_{T_A});\ T_A < \infty],$$

where

$$T_A = \inf\{t > 0\ ;\ X_t \in A\}$$

denotes the hitting time of the set A; here, f is a Borel measurable function. All other notations used throughout this paper are generally those of Blumenthal - Getoor [1].

Now we have the following

Proposition 1. All excessive functions of the process X are locally integrable if and only if

(1) $$\int u(x,y)g(x)dx$$

is bounded as a function of y for all bounded functions g with compact support.

Proof. Suppose excessive functions of X are locally integrable. Let K be a compact set. Each finite measure m on K determines an excessive function by the prescription

(2) $$s(x) = \int u(x,y)m(dy)$$

because for x not in K, s(x) is finite. Function s is integrable on K.

Since this holds for all choices of finite measures m on K, $\int_K u(x,y)dx$ must be bounded on K and hence everywhere by the maximum principle. See [7]. Conversely, if s is excessive and D a relatively compact open set, s agrees with $P_D s$ on D and the Riesz measure of $P_D s$ is a finite measure concentrated on the closure of D. If (1) holds, the integrability of s on D follows immediately by the Fubini theorem.

Theorem 2. Suppose excessive functions are locally integrable and the transition probabilities have densities p(t,x,y) which are bounded and continuous in x,y, the bounds depending on t but not on x,y. Then $P_t s$ is continuous for all excessive functions s which are not identically infinite.

Proof. The proof of this theorem will be given in five steps.

Step 1. $P_t s$ is everywhere finite. The proof of this fact is exactly as in the Brownian motion case. See pp. 5.44 of [8]. We need here the finiteness of $P_{D^c} s$ which is guaranted by Theorem 3. of [4].

Step 2. If s is excessive and $P_t s = s$ then s is harmonic, hence continuous by [4].

Indeed, write s = p + h, with p potential and h harmonic. Then

$$P_t p = p. \tag{3}$$

If m is the Riesz measure of p, m is a Radon measure so (3) implies that for each fixed x, $P_t u(x,y) = u(x,y)$ for m-almost all y. Hence, by Fubini theorem $P_t u(x,y) = u(x,y)$ for m-almost all y and almost all x. However, by excessivity of both sides, the above statement is true for m-almost all y and all x. On the other hand, $u(y,y) = \infty$, whereas from Step 1, $P_t u(y,y) < \infty$. Hence, m must be zero.

Step 3. Given an excessive function s. We can write s = g + h where g is purely excessive, h excessive and satisfying $h = P_t h$, hence harmonic by the conclusion of the Step 2, and, therefore continuous by [4].

Step 4. Given an excessive function s. By the conclusion of the Step 3 without the loss of generality we may assume that s is purely excessive. Put $v = P_t s$. Claim: as D increases to the state space E, $P_{D^c} v$ decreases to zero uniformly on compacts.

Indeed, for any positive integer n, if $T = T_{D^c}$, one has:

$$E^x[s(X_{T+t})] = E^x[s(X_{T+t}):T< nt] + E^x[s(X_{T+t}):T\geq nt]$$
$$\leq E^x[s(X_{T+t}):T< nt] + E^x[s(X_{nt})].$$

Since s is purely excessive the second term on the right-hand side of this inequality tends to zero as $n\to\infty$. Choose n sufficiently large so that this term is small. For such n, one gets the following estimate for the first term on the right-hand side of the last inequality:

$$E^x[s(X_{T+t}):T<nt] = \sum_{m=0}^{n-1} E^x[s(X_{T+t}):mt\leq T< (m+1)t]$$
$$\leq \sum_{m=0}^{n-1} E^x[s(X_{(m+1)t}):T< (m+1)t] .$$

Since $P_t s(x)<\infty$, it follows that $P_{mt}s(x)<\infty$ for all m, hence, as D increases to E each summand of the finite sum on the right-hand side of the last inequality tends to zero. Thus $P_{D^c} v$ decreases to zero as D increases to E. Now one appeals to Dini's Theorem which can be applied here due to Theorem 3 of [4].

Step 5. Since $P_{2t}s = P_tP_ts$, we may assume by Step 4 that $P_{D^c}s$ decreases to zero uniformly on compact as D increases to E. If $T = T_{D^c}$, we have:

$$P_t s(x) = E^x[s(X_t):t< T] + E^x[s(X_t):T\leq t]. \tag{4}$$

The last term above is dominated by $P_{D^c}s$ which is uniformly small on compacts as D increases to E. It remains to show that the first term on the right-hand side (4) is continuous. Denoting by Q_t the transition semi-group of the process killed upon exit from D, i.e., $Q_tf = E^x[f(X_t):T> t]$, the second term in (4) is just $Q_ts(x)$. Since

P_t has bounded densities so does Q_t. Because s is integrable in D, $Q_{\frac{t}{2}}s$ is bounded in D. Now P_t is clearly strong Feller semi-group. But, so is Q_t. Here is the proof. First, it is well known that for each fixed $t > 0$, the function in x:

$$P^x[T > t] \tag{5}$$

is upper semi-continuous. See [6,p. 11]. It is also excessive relative to the Q_t defined above. Second, it has been shown in [5] that the killed process satisfies the hypothesis of the original process, so that all excessive functions relative to Q_t are necessarily lower semi-continuous. Hence, the function in (5) is continuous in D.

Now, for any measurable function f satisfying $0 \le f \le 1$ write

$$P_h[Q_{t-h} f] = Q_t f + E^x[Q_{t-h} f(X_h) ; T \le h]. \tag{6}$$

The last term in (6) is less than or equal to $P^x[T \le h]$. As $h \downarrow 0$ these continuous functions decrease to zero. By Dini's Theorem, the convergence is uniform on compact subsets of D. Therefore, (6) implies that continuous functions $P_h[Q_{t-h} f]$ converge uniformly on compact subsets of D to $Q_t f$. So, Q_t is strong Feller, as asserted.

Since $Q_{\frac{t}{2}} s$ is bounded in D, $Q_t s = Q_{\frac{t}{2}}(Q_{\frac{t}{2}} s)$ is continuous in D. That completes the proof.

<u>Remark.</u> In addition to the fact that semi-group Q_t is strong Feller, one can show much more: namely, there exists a function $q(t,x,y)$ defined for all x,y and $t > 0$ such that

$$\int q(t,x,z) q(s,z,y)\, dz = q(t+s,x,y) ,$$

where $q(t,.,.)$ is upper semi-continuous in (.,.), $q(t,.,y)$ is a continuous function on D, and for every $t > 0$ and $x \in D$, $q(t,x,.)$ is a density for $Q_t(x,.)$.

To see this, one starts from the first passage time relation:

$$P_t f(x) = Q_t f(x) + E^x[f(X_t) ; T < t].$$

Now, let us define a function $q(t,x,y)$ for all x,y and $t > 0$ by the relation:

(7) $$p(t,x,y) = q(t,x,y) + E^x[p(t-T,X_T,y):T<t],$$

where p(t,x,y) is a density of $P_t(x,.)$. We shall show that this function q is the desired density. First, it is clear from (7) that q(t,x,.) is a density of $Q_t(x,.)$ for each x. Second, by using the semi-group property of p(t,.,.) one gets:

(8) $$p(t,x,y) = \int q(t-\varepsilon,x,z)p(\varepsilon,z,y)\,dz + E^x[p(t-T,X_T,y):T<t-\varepsilon],$$

with $\varepsilon > 0$. By comparing (7) and (8), one obtains:

$$\int q(t-\varepsilon,x,z)p(\varepsilon,z,y)\,dz = q(t,x,y) + E^x[p(t-T,X_T,y):t-\varepsilon \le T<t].$$

In particular,

(9) $$\int q(t-\varepsilon,x,z)p(\varepsilon,z,y)dz \downarrow q(t,x,y) \text{ as } \varepsilon \downarrow 0, \text{ for all } x,y.$$

By Markov property, $Q_t(x,.)$ is a semi-group of measures. In terms of densities this means: for every x and every t,s > 0, and for almost all y,

(10) $$q(t+s,x,y) = \int q(t,x,z)q(s,z,y)\,dz.$$

But by (9), (10) must hold for all y.

Finally, it remains to show that q(t,x,y) is upper semi-continuous in (x,y). To see this, observe that $s + T(\theta_s) \downarrow T$ as $s \downarrow 0$. Then, it follows that for every nonnegative, measurable function f,

(11) $$\begin{aligned} E^x[f(X_t):t<T] &= \lim_{s\downarrow 0}\downarrow E^x[f(X_t):t<s+T(\theta_s)] \\ &= \lim_{s\downarrow 0}\downarrow \int p(s,x,z)Q_{t-s}f(z)\,dz. \end{aligned}$$

Now for s = t, (10) gives $q(2t,x,y) = Q_t(q(t,.,y))(x)$. Using this fact, (11) with f = q(t,.,y) implies that:

$$q(2t,x,y) = \lim_{s\downarrow 0}\downarrow \int p(s,x,z)Q_{t-s}(q(t,.,y))(z)\,dz.$$

However, from (9) it follows that:

$$q(t,\xi,y) = \lim_{\varepsilon\downarrow 0}\downarrow \int p(\varepsilon,z,y)q(t-\varepsilon,\xi,z)\,dz.$$

Therefore, one has:

$$q(2t,x,y) = \lim_{\varepsilon\downarrow 0}\downarrow \int p(\varepsilon,x,z)q(t-\varepsilon,z,\xi)q(t-\varepsilon,\xi,\eta)p(\varepsilon,\eta,y)\,d\xi d\eta dz.$$

For every $\varepsilon > 0$, the expression under the limit is clearly continuous in (x,y), thus showing that q(t,x,y) is upper semi-continuous in (x,y). Finally, the fact that q(t,x,y) is continuous in x for each y is a consequence of the semi-group property and the strong Feller property of Q_t.

ACKNOWLEDGMENT. The author wishes to express his profound gratitude to Professor K. Murali Rao for valuable suggestions concerning this paper.

REFERENCES

[1] R.M. Blumenthal and R.K. Getoor, Markov Processes and Potential Theory, New York, Academic Press, 1968.

[2] K.L. Chung and M. Rao, A new setting for Potential Theory, Ann. Inst. Fourier 30, 1980, 167-198.

[3] K.L. Chung, Probabilistic approach in Potential Theory to the equilibrium problem, Ann. Inst. Fourier 23, 1973.

[4] Z.R. Pop-Stojanovic, Continuity of Excessive Harmonic Functions for certain Diffusions, Proc. of the AMS, V. 103, N. 3, 1988.

[5] Z.R. Pop-Stojanovic, Excessiveness of Harmonic Functions for certain Diffusions, Pre-print, 1988.

[6] S. Port and C. Stone, Brownian Motion and classical Potential Theory, New York, Academic Press, 1978.

[7] Murali Rao, A note on Revuz Measure, Seminaire de Probabilites XIV, 1978/79, Lecture Notes in Mathematics # 784, Springer-Verlag, New York-Heidelberg, 1980, 418-436.

[8] Murali Rao, Brownian Motion and Classical Potential Theory, Lecture Notes Series # 67, 1977, Aarhus University, Denmark.

Department of Mathematics
University of Florida
Gainesville, Florida 32611

A MAXIMAL INEQUALITY

by

K. Murali Rao

Let X be a uniformly integrable, cadlag non-negative regular supermartingale. Such a process X has the representation

(1) $$X_t = E[A_{\infty+} - A_t | F_t]$$

where A_t is continuous and increasing on the half open interval $[0,\infty)$, $A_0 = 0$ and A may assign mass to ∞ which is just $A_{\infty+} - A_\infty$ where $A_\infty = \lim_{t\uparrow\infty} A_t$. Then we have the maximal inequality.

<u>Theorem 1</u> . Let λ,μ be non-negative numbers. Then

(2) $$\mu P[X^* \geq \lambda + \mu] \leq E[\int_T^\infty 1_{X_u > \lambda} dA_u ; T < \infty] \leq E[\int_0^\infty 1_{X_u \geq \lambda} dA_u ; X^* \geq \lambda + \mu]$$

where

$$T = \inf\{t ; X_t \geq \lambda + \mu\} .$$

Before the proof note that if $\lambda \to 0$ in (2) we get the standard inequality

$$\mu P[X^* \geq \mu] \leq E[A_\infty ; X^* \geq \mu] .$$

<u>Proof</u> . Let $\lambda > 0$ be fixed. First we prove that

(3) $$X_t \leq \lambda + E[\int_{[t,\infty]} 1_{X_u > \lambda} dA_u | F_t] .$$

Because of (1), (3) is equivalent to

(4) $$E[\int_{[t,\infty]} 1_{X_u \geq \lambda} dA_u | F_t] \leq \lambda.$$

Define the stopping time R by

$$R = \begin{cases} \inf\{u ; u \geq t, X_u \leq \lambda\} , \\ \infty \quad , \text{if there is no such } u . \end{cases}$$

Then

(5) $$\int_{[t,\infty]} 1_{X_u \le \lambda}\, dA_u = \int_{[R,\infty]} 1_{X_u \le \lambda}\, dA_u \le A_{\infty+} - A_R .$$

Here we are using the continuity of A; otherwise we would have to write at the end $A_{\infty+} - A_{R-}$. Now since $R \ge t$ the conditioning in (4) may first be done on F_R and so using (5) and the definition of X we see that

$$E[\int_{[t,\infty]} 1_{X_u \le \lambda}\, dA_u | F_R] \le E[X_t | F_R] \le \lambda ,$$

establishing (3). Write

(6) $$\bar{Y}_t = E[\int_{[t,\infty]} 1_{X_u \ge \lambda}\, dA_u | F_t] .$$

$\bar{Y}_t$ is cadlag supermartingale and we may write (3) as

(7) $$X_t \le \lambda + \bar{Y}_t.$$

($\bar{Y}$ depends on λ). Let μ be any positive number and let

$$T = \inf\{t; X_t > \lambda + \mu\}.$$

From (7) we get

(8) $$X_T 1_{T<\infty} \le \lambda 1_{T<\infty} + \bar{Y}_T 1_{T<\infty} .$$

Thus if $X^* = \sup_t X_t$,

$$(\lambda+\mu)P[X^* \ge \lambda + \mu] \le E[X_T ; T<\infty] \le \lambda P[T<\infty] + E[\bar{Y}_T ; T<\infty]$$

or that

$$\mu P[T<\infty] \le E[\bar{Y}_T ; T<\infty]$$

which is (2). Q.E.D.

[Recall $X_\infty = \lim_{t\uparrow\infty} X_t$ so that $X_\infty > \lambda + \mu$ implies $X_t > \lambda + \mu$ for $t<\infty$] .

Corollary 2 . Let $p > 1$. Then for $\lambda > 0$,

(9) $$||(X^*-\lambda)^+||_p \le q\, E[(\int_0^\infty 1_{X_u \ge \lambda}\, dA_u)^p ; X^* \ge \lambda]^{1/p}.$$

Proof . Write (2) as

$$\mu P[(X^*-\lambda)^+ \ge \mu] \le E[B ; (X^*-\lambda)^+ \ge \mu]$$

where $B = \int_0^\infty 1_{X_u \geq \lambda} dA_u$. Standard argument then gives (9). For $\lambda > 0$ we get the standard

$$||\dot{X}^*||_p \leq q||A_\infty||_p .$$

<u>Corollary 3</u>. The following inequality holds:

$$E[X^*] \leq 2 + 2E[\int_0^\infty \log^+ X_u \, dA_u] . \tag{10}$$

<u>Proof</u>. Take $\mu = \lambda$ in (2) to get

$$P[X^* \geq 2\lambda] \leq \lambda^{-1} E[\int_0^\infty 1_{X_u > \lambda} dA_u ; X^* \geq 2\lambda] .$$

Then

$$E[X^*] = 2\int_0^\infty P[X^* \geq 2\lambda]\, d\lambda \leq 2 + 2\int_1^\infty P[X^* \geq 2\lambda]\, d\lambda$$

$$\leq 2 + 2E[\int_0^\infty dA_u \int_1^\infty 1_{X_u > \lambda} 1_{X^* \geq 2\lambda} \lambda^{-1} d\lambda]$$

$$\leq 2 + 2E[\int_0^\infty \log^+ X_u \, dA_u] . \quad \text{Q.E.D.}$$

Almost the same argument as in the proof of Theorem 1 gives the following:

Let N_t be a non-negative supermartingale. Then

$$X_t \leq N_t + E[\int_{[t,\infty]} 1_{X_u \geq N_u} dA_u | F_t] . \tag{11}$$

In particular take for a <u>given s</u> $N_t = E[A_s | F_t]$. From (11) we get for $t = s$

$$X_s \leq A_s + E[\int_{[s,\infty]} 1_{X_u \geq A_s} dA_u | F_s] . \tag{12}$$

We have the following corollary which seems better than Corollary 2.

<u>Corollary 4</u>. Let $p > 0$ and $\mu > 0$. Then

$$p\,\mu E[(X^* - \mu)^p] \leq E[\int_0^\infty X_u^p \, dA_u] \leq E[(X^*)^p A_\infty] . \tag{13}$$

<u>Proof</u>. Indeed write (2) in form

$$\mu P[(X^* - \mu)^+ \geq \lambda] \leq E[\int_0^\infty 1_{X_u \geq \lambda} dA_u ; (X^* - \mu)^+ \geq \lambda] \leq E[\int_0^\infty 1_{X_u \geq \lambda} dA_u] .$$

Multiply both sides by $p\lambda^{p-1}$ and integrate from 0 to ∞ to get (13).
Q.E.D.

References

[1] C. Dellacherie and P.A. Meyer, Probabilites et Potentiel, Chapitres V a VIII, Hermann, Paris, (1980).

Department of Mathematics
University of Florida
Gainesville, Florida 32611

SOME RESULTS FOR FUNCTIONS OF KATO CLASS IN DOMAINS OF INFINITE MEASURE

By

Murali Rao

<u>Introduction</u>. In this note we extend the gauge theorem to certain sets of infinite measure provided they are "Small" at infinity. For such sets when the gauge is bounded we show that the Schrödinger-Green Kernel is weakly compact in L' and compact in L^p for $1<p<\infty$.

The first of the following results is in response to a problem raised by K. L. Chung.

Let Ω be a connected open subset of R^d, $d>3$. We denote by G the green function of Ω - this is the inverse of $-\Delta$ (the Laplacian) with Dirichlet boundary conditions. Let

(1) $K(x) = |x|^{-d+2}$, $d>3$.

A Borel function q defined on R^d is said to belong to the Kato class K_d if

(2) $\lim_{\alpha\to 0} \{\sup_x \int_{|x-y|<\alpha} K(x-y)|q|(y)dy = 0.$

Let X denote the Brownian motion in R^d.

Put

(3) $e(t) = \exp [\int_0^t q(X_s)ds].$

The function

(4) $g(x) = E^x[e(\tau)]$

where τ is the exit time from Ω has been called the Gauge of Ω. by K. L. Chung. When Ω has finite measure it is known [1] that finiteness of g at one point implies that g is bounded and continuous in Ω. This result has been called the gauge theorem by K. L. Chung. It is of interest to know, for which domains of infinite measure, the result remains valid. In these considerations, boundedness of $G|q|$ plays an important role. K. L. Chung wants to know (private communication) the necessary and sufficient condition on the domain guaranteeing the boundedness of $G|q|$ for q in the Kato class. We are only able to provide a very general sufficient condition. This same condition was used in [2] to obtain a lower bound for the premier eigen-value of the Laplacian.

Fix $p>0$ and $0<\varepsilon<1$. We say that a domain Ω satisfies the (p,ε) condition if in each ball $B(x,p)$ with centre x and radius p, the (Newlonian) capacity of the complement of Ω in $B(x,p)$ is at least εp^{d-2}.

Let q be a Borel function on R^d and put

$$(5)\qquad M_R = \sup_x \int_{|x-y|<R} K(x-y)|q|(y)dy.$$

With these definitions we have the following theorem.

Theorem 1. Let Ω satisfy a (p,ε) condition. Let q be Borel such that for some $R>0$, and hence all R, M_R defined in (5) is finite. Then $G|q| < \frac{2M}{\varepsilon}$.

Proof. We assume $q>0$. A simple compactness argument implies that M_R is finite for all R, if it is finite for one. Recalling p from the statement of the theorem put

$$S_1 = \inf \{t = |X_t - X_0| > 2^{\frac{1}{d-2}} p\}$$

and

$$S_{n+1} = S_n + S_1(\theta_{S_n}),\ n>1.$$

From (5)

$$(6)\qquad \sup_y E^y[\int_0^{S_1} q(X_s)ds] < M < \infty.$$

Let τ = exit time from Ω. Following the arguments in [2] it is seen that

$$(7)\qquad \sup_x p^x[S_1<\tau] < 1-\frac{\varepsilon}{2}.$$

In particular $P^x[\tau<\infty] = 1$ for all x.

For any x

$$Gq(x) = E^x[\int_0^\tau q(X_s)ds] = E^x[\int_0^\tau q(X_s)ds:\ \tau<S]$$

$$+ \sum_1^\infty E^x[\int_{S_n}^\tau q(X_s)ds:\ S_n<\tau<S_{n+1}]$$

Using standard arguments we get from (7)

$$< M\ [1+ \sum_1^\infty (1-\frac{\varepsilon}{2})^n] = \frac{2M}{\varepsilon}.$$

Completing the proof.

Remark. We did not assume that q belonged to the Kato class in the above theorem. Clearly a domain with finite measure has the property stated in the theorem for some p and $\frac{1}{2} = \varepsilon$. It is simple to obtain many examples of domains satisfying the conditions of the theorem and which are of infinite measure.

If q belongs to the Kato class the following result holds.

Corollary 2 Let q belong to the Kato class and Ω satisfy the (p,ε) condition for some p,ε. Then $G|q|$ is bounded and continuous in Ω.

For the proof note that if B is a ball contained in Ω, $G_B|q|$ is continuous, this is standard. G_B is the Green function of B. And we have

$$G|q| = G_B|q| + h$$

where h is harmonic in B with boundary values $G|q|$. The continuity of $G|q|$ follows.

If q belongs to the Kato class, M_R tends to zero as $R \to 0$. Thus if p is small enough say and $M_{2p} < \frac{1}{6}$ and a domain Ω satisfies, $(p, \frac{1}{2})$ condition we get

$$\sup_x G|q|(x) < \frac{2}{3} < 1.$$

The following statement is thus immediate.

<u>Theorem</u> 3. Let Ω satisfy the $(p, \frac{1}{2})$ condition for small enough p. (p such that say $M_{2p} < \frac{1}{6}$). Then gauge is bounded i.e.

$$\sup_x E^x[e(\tau)] < \frac{1}{1-\frac{2}{3}} = 3.$$

<u>Remark</u>. Any open set of finite measure satisfies the $(p, \frac{1}{2})$ condition for some p. If the set has small enough measure, p can also be chosen small. Any cylinder of small radius satisfies a $(p, \frac{1}{2})$ condition for suitably small p but may have infinite measure. Theorem 3 tells us that the gauge of a cylinder of small enough radius is bounded.

<u>The gauge theorem</u>.

Let q belong to the Kato class and Ω connected open of finite measure. Then it is proved in [1] that the gauge is bounded and continuous in Ω unless it is identically infinite. K. L. Chung calls it the gauge theorem. It is of interest to find domains not necessarily of finite measure for which the gauge theorem still holds.

We make the following definition.

Definition. An open connected set Ω will be called small at infinity if given $\varepsilon>0$ we can find a compact set K such that Ω satisfies the $(p, \frac{1}{2})$ condition for some $p<\varepsilon$ at all points outside K. In other words, the capacity of the complement of Ω in each ball of radius p, whose centre is not in K, is at least $\frac{1}{2} p^{d-2}$.

With this definition we have

Theorem 4. Let Ω be a connected open set which is small at infinity. Then the gauge is bounded and continuous unless it is identically infinite.

Proof. First choose $\varepsilon>0$ such that $M_{2\varepsilon} < \frac{1}{16}$ whose M_R is defined in (5). This is possible because q belongs to the Kato class. Next choose a compact set K such that Ω satisfies the $(p, \frac{1}{2})$ condition at all points outside K, where $p<\varepsilon$. Put

$$S_1 = \inf \{t: |X_t - X_0| > 2p\}.$$

Then if $x \notin K$,

$$(8) \qquad P^x[S_1 < \tau] < 1 - \frac{1}{4} = \frac{3}{4}.$$

Now suppose g is not identically infinite. Then it is easy to see that the set $(g = \infty)$ is polar. In particular it has measure zero. g is l.s.c. [1]. For each n, $(g<n)$ is therefore closed. For N large enough then, if η is the exit time from the set $U = K \cap \Omega \cap (g>N)$

$$(9) \qquad P^x[S_1 < \eta] < \frac{1}{16}, \quad x \in U$$

Let T be the exit time from the set $W = (g>N)$.

$= U \cup (g>N) \cap K^C$

We have if $x \in (g>N) \cap K^C$ from (8)

$$P^x[S_1 < T] < P^x[S_1 < \tau] < \frac{3}{4}$$

If $x \in U$ then from (9) and (8)

$$P^x[S_1 < T] = P^x[S_1 < \eta] + P^x[\eta < S_1 < T]$$

$$< \frac{1}{16} + E^x[P^{X_\eta}\{S_1 < T\}: \eta < S_1, \eta < T]$$

$$< \frac{1}{16} + \frac{3}{4} < \frac{13}{16}.$$

Thus for all $x \in \Omega \setminus K$,

$$P^x[S_1 < T] < \frac{7}{8}.$$

As in Theorem 1 we get for all $x \in \Omega \setminus K$

$$E^x[\int_0^T |q|(X_s)ds] < \frac{1}{2}.$$

So that

$$E^x[e(T)] < 3, \quad x \in \Omega \setminus K.$$

Now the proof is exactly as in Theorem 1 of [1].

Remark. Many other results of [1] carry over to this case. In particular the super gauge theorem remains valid.

The Schrödinger - Green Kernel.

Let Ω be small at infinity and assume that the gauge is bounded and continuous in Ω.

For $f > 0$ put

$$Kf(x) = E^x[\int_0^\tau e(t)f(X_t)dt].$$

(10) $\quad g(x) = E^x[e(\tau)].$

The following proposition connects the green Kernel with the gauge. Same letter may signify different constants.

<u>Proposition</u> 5. For a constant $m>0$.

(11) $m\, K|q|(x) < g(x) < [1+K|q|(x)]$

<u>Proof</u>. Since

$$e(\tau) = 1+ \int_0^\tau q(X_s)e(s)ds < 1+ \int_0^\tau |q|(X_s)e(s)ds$$

the last half of (11) is immediate.

Since 1 belongs to the Kato class $\sigma = G1$ is bounded. So for large N, $P^x[\tau>N]$ is uniformly small. By Jensen's inequality

$$E^x[e(\tau)\colon \tau<N] > P^x[\tau<N]\exp[- \frac{1}{P^x[\tau<N]} E^x[\int_0^\tau |q|(X_t)dt\colon \tau<N]]$$

$$> P^x[\tau<N]\exp[- \frac{1}{P^x[\tau<N]} G|q|(x)] > m .$$

Then

$$g(x) = \sum_0^\infty E^x[e(\tau)\colon nN<\tau<(n+1)N]$$

(12) $$> \sum_n E^x[e(nN)E^{X_{nN}}[e(\tau)\colon \tau<N]\colon \tau>nN]$$

$$>m \sum_0^\infty E^x[e(nN)\colon \tau>nN]$$

And

$$K|q|(x) = \sum_0^\infty E^x[\int_{nN}^{(n+1)N\wedge\tau} |q|(X_t)e(t)dt\colon \tau>nN]$$

(13) $$\le \sum_0^\infty E^x[e(nN)E^{X_{nN}}\{\int_0^\tau |q|(X_t)e(t)dt:\tau<N\}:\tau>nN]$$

$$\le A \sum_0^\infty E^x[e(nN):\ \tau>nN]$$

where $A = \sup_x E^x[\int_0^N |q|(X_t)e(t)dt]$

(12) and (13) give us (11).

We are assuming that the gauge is bounded. Proposition 5 permits the use of Fubini's theorem and we have, if f is bounded

$$K\ f = G\ f + K[qGf]$$

(14) $$K\ f = G\ f + G[qKf].$$

Now G is symmetric and hence so is K. K is an integral kernel: this follows from general principles. G1 is bounded, $K|q|$ is bounded and therefore from the first relation in (14) K 1 is bounded. Let $0 \le f \in L^p$. Using Hölder

$$(Kf(x))^p = (\int K(x,y)f(y)dy)^p \le K1(x)^{\frac{p}{q}} \int K(x,y)f^p(y)dy$$

$$\le \|K1\|_\infty^{p/q}\ Kf^p(x)$$

Integration of both sides shows that K maps $L^p \to L^p$ with norm at most $\|K1\|_\infty$.

Next we want to prove that K is weakly compact on L' and compact on L^p for $1<p<\infty$. In preparation for this we have

Proposition 6. Given $\eta>0$, there is a bounded subject F of Ω such that if ϕ is the indicator of the complement of F in Ω then

(15) $\quad K\phi < \eta$

Proof. Let F_1 be a bounded set such that Ω satisfies the $(p, \frac{1}{2})$ condition for some $p < \varepsilon$, ε will be determined later. If S_1 is as in Theorem 4 then

$$P^x[S_1<\tau] < \frac{3}{4}, \quad x \notin F_1$$

Since 1 belongs to the Kato class we know that

$$\sigma(x) = G1(x)$$

is bounded and continuous in Ω. Choose integer (smallest) n such that

(16) $\quad (\frac{3}{4})^n \|\sigma\|_\infty < \varepsilon.$

and let F be the (bounded) set of y whose distance from F_1 is at most 2np. For any $x \notin F$,

$$\sigma(x) < E^x[S_n: \tau<S_n] + E^x[\sigma(X_{S_n}): S_n<\tau]$$

$$< n\, E^x[S_1] + \varepsilon = n\,\frac{(2p)^2}{d} + \varepsilon < \frac{4\varepsilon^2 n}{d} + \varepsilon$$

$$< n\varepsilon \text{ if } \varepsilon < \frac{1}{4}.$$

because the distance of x from F_1 being at least 2np, X_{S_j} is almost surely in F_1^c for $j < n$.

We have proved that for $x \notin F$,

$$\sigma(x) < n\varepsilon.$$

Since n and ε are connected by (16), $n\varepsilon \to 0$ as $\varepsilon \to 0$. In particular $\sigma(x) \to 0$ as $|x| \to \infty$.

If ϕ is the indicator of $\Omega \setminus F$, $G\phi < \sigma(x) < n\varepsilon$ for $x \in \Omega \setminus F$ and hence for all x by the maximum principle.

From the first relation in (14) with f replaced by ϕ we get

$$K\phi < n\varepsilon[1+\|K|q|\|_\infty.]$$

Since $n\varepsilon \to 0$ we can choose ε to satisfy (15) as claimed.

Proposition 7. Let B be a measurable set. And ϕ its indicator function. Then

$$K\phi < [1+\|K|q|\|_\infty]w|B|^{2|d}.$$

where $|B|$ = measure of B and w the area of the unit sphere. It is simple to show that

$$G\phi < w|B|^{2/d}.$$

See for example the proof of Theorem 1 in [2]. Rest follows as in Proposition 6, from the first identity in (14).

New we can prove the weak compactness in L' of K. Indeed let f_n be bounded in L'. We know then that $F_n = Kf_n$ is also bounded in L'. Given $\varepsilon>0$ we can find according to Proposition 6, a bounded set A such that $\int_{A^c} K(x,y)dy < \varepsilon$. Then $\int_{A^c} Kf_n < \varepsilon\int f_n$ is also uniformly small. Similarly by Proposition 7, the integral of Kf_n over a set of small measure is also uniformly small. In other words the set $\{Kf_n\}$ is relatively weakly compact in L^1.

Now consider L^p, $1<p<\infty$. Let $0<f_n$ be a sequence which converges weakly to f. We know $K1(x) = \int K(x,y)dy$ is bounded. Given x, for each n, $K(x,y) \wedge n$ is then L^q for all q. So

$$\int (K(x,y)\wedge n)f_m \to \int (K(x,y)\wedge n)f.$$

Letting n tend to infinity we conclude

(17) $\liminf Kf_n \geq Kf.$

Let ϕ be bounded continuous strictly positive and integrable in Ω. Then $K\phi$ is bounded and integrable and hence in L^p for all p. Since $f_n \to f$ weakly in L^p we have

$$\lim \int \phi Kf_n = \lim \int f_n K\phi = \int fK\phi = \int \phi Kf$$

Together with (17) this tells us that Kf_n tends to Kf in ϕ-measure and hence in measure at least on bounded subsets of Ω. The same is then true of $(Kf_n)^p$. This sequence is also uniformly integrable because

$$(Kf_n)^p \leq \|K1\|_\infty Kf_n^{\,p}.$$

f_n^p is bounded in L' and so previously proved result applies. Thus

$$\lim_n \int (Kf_n)^p = \int (Kf)^p.$$

Since we already know Kf_n converges to Kf weakly in L^p the above tells us that it does so strongly in L^p.

We gather all the above in

<u>Theorem</u> 8. Let Ω be small at infinity. Suppose its gauge is bounded. Then K is a weakly compact operator in L' and strongly compact in L^p for $1<p<\infty$.

REFERENCES

[1] K. L. Chung and K. M. Rao, General Gauge Theorem for Multiplicative Functionals. Trans. A.M.S., 306 (1988) 819-836.

[2] P. J. McKenna and Murali Rao, Lower Bounds for the First Eigen Value of Laplacian. Applicable Analysis 18 (1984) 55-66.

Murali K. Rao
Department of Mathematics
University of Florida
201 Walker Hall
Gainesville, FL 32611

SOME PROPERTIES OF INVARIANT FUNCTIONS OF MARKOV PROCESSES

by

WU RONG

Let $X = (\Omega, F, F_t, X_t, \theta_t, P^x)$ be a right process on a Lusin topological state space E with Borel field B. A point $\Delta \in E$ will serve as cemetery point. Let P_t and U^a denote the semigroup and resolvent of X. We suppose X is a Borel right process; in particular, $U^a f \in B^+$ whenever $f \in B^+$. We restrict our attention to transient Borel right processes throughout this paper, so there is a strictly positive B-measurable function q so that $Uq \leq 1$.

Let $\hat{X} = (\hat{\Omega}, \hat{F}, \hat{F}_t, \hat{X}_t, \hat{\theta}_t, \hat{P}^x)$ be another transient Borel right process on (E, B) with semigroup $\hat{P}_t$ and resolvent $\hat{U}^a$. Let m be a fixed sigma-finite measure on (E, B). If

(i) $\int_E P_t f \cdot g \, dm = \int_E f \cdot \hat{P}_t g \, dm$ for all $f, g \in B^+$;

(1) (ii) X_{t-} exists for all t in $[0, \zeta)$ a.s.;

(iii) $\hat{X}_{t-}$ exists for all t in $[0, \hat{\zeta})$ a.s.,

then we say X and $\hat{X}$ are in weak duality with respect to m (or the triple $(X, \hat{X}, m)$ is in weak duality). The left hand side of (1 i) will be denoted by $(P_t f, g)$.

Let S^a (resp. $\hat{S}^a$) be the collection of B-measurable functions which are a-excessive for X (resp. $\hat{X}$) and which are finite m a.e. Let μ be a sigma-finite positive measure on (E, B) and define an a-excessive measure by setting

(2) $$\mu U^a = \int \mu(dx)\, U^a(x,\cdot)$$

Let M^a (resp. M^a) be the class of sigma-finite measures μ on (E, B) not charging cofinely open m-copolar sets with $\mu\hat{U}^a$ sigma-finite (resp. finely open m-polar sets with μU^a sigma-finite).

DEFINITION 1. A function $f \in S^a$ will be called the a-potential of a measure μ if $fm = \mu\hat{U}^a$.

DEFINITION 2. A function $f \in B^+$ will be called an invariant function of X if for all t and x,

(3) $$f(x) = P_t f(x) .$$

For the definition of harmonic function, we take the one given in (4.3) of [3].

THEOREM 1. Suppose that $(X, \hat{X}, m)$ is in weak duality and f is an invariant function of X which is finite m a.e. Then there exists a harmonic function h of X such that

(4) $$f = h \quad \text{a.e. } m .$$

PROOF. Since $f \in S$, by applying the result of Theorem (4.5) in [3] to f, we have

(5) $$f(x) = U\mu(x) + h(x) \quad \text{m a.e.}$$

where $\mu \in M$ and h is a harmonic function of X. Moreover, μ is unique and h is unique a.e. m. First we prove

(6) $$\lim_{t\to\infty} P_t U\mu = 0 \quad \text{m a.e.}$$

Since $\mu\hat{U}$ is sigma-finite, there is a function $g \in B^+$ such that $\mu\hat{U}(g) = (g, U\mu) < \infty$ and

$$(g, P_t U\mu) = (U\mu, \hat{P}_t g) = \mu\hat{U}(\hat{P}_t g) = \int \mu(dx) E^x \int_t^\infty g(X_s)\,ds \le \mu\hat{U}g < \infty$$

Using the dominated convergence theorem, we obtain $\lim_{t\to\infty}(g, P_t U\mu) = 0$, and since $U\mu\cdot m$ is sigma-finite, we have

$$\lim_{t\to\infty} P_t U\mu = 0 \quad \text{m a.e.}$$

Since $f \in S$, there is a function $g \in B^+$ which is bounded such that $(f,g) < \infty$. From (5) then,

$$(f, \hat{P}_t g) = (g, P_t f) = (g, P_t U\mu) + (g, P_t h) .$$

By the result above and (6), we have $f = \lim_{t\to\infty} P_t h$ m a.e. Since $h \in S$, we have

$$f = \lim_{t\to\infty} P_t h \leq h \leq f \quad \text{m a.e.}$$

Hence $f = h$ m a.e. Q.E.D.

DEFINITION 3. A sigma-finite measure μ will be called an invariant measure if for every $t > 0$,

(8) $$\mu = \mu P_t .$$

THEOREM 2. Suppose that $(X, \hat{X}, m)$ is in weak duality.

(i) If f is an invariant function of X which is finite m a.e., then there is an invariant measure $\hat{\mu}$ of $\hat{X}$, and $\hat{\mu} = f\cdot m$.

(ii) If μ is a sigma-finite measure which does not charge cofinely open m-copolar sets and which is invariant for X, then it is absolutely continuous with respect to m, and we can choose $f_\mu = d\mu/dm$ such that $f_\mu \in \hat{S}$ and

(9) $$f_\mu(x) = P_t f_\mu(x) \quad \text{m a.e.}$$

PROOF. (i) Since f is finite m a.e., $f\cdot m$ is a sigma-finite measure. We need only show that $f\cdot m$ is an invariant measure of X. If $g \in B^+$ and $\hat{\mu} = f\cdot m$, then $\hat{\mu}(g) = (f,g) = (P_t f, g) = (f, \hat{P}_t g) = \hat{\mu}(\hat{P}_t g)$. This means that $\hat{\mu}$ is an invariant measure of $\hat{X}$.

(ii) Since $\mu = \mu P_t$,

(10) $$\mu = a\mu U^a$$

for every $a > 0$. It is immediate from (10) that $\mu \in \hat{M}^a$. From Proposition (3.8) in [3], $\mu U^a \ll m$, and there is an a-potential $\hat{U}^a\mu \in \hat{S}^a$ such that $\mu U^a = \hat{U}^a\mu \cdot m$. By (10), it is easy to check that $\mu \ll m$.

From Proposition (3.6) in [3], we can choose $f_\mu = d\mu/dm \in \hat{S}$. Using (10) again, for any $g \in B^+$, we have

(11) $$\mu(g) = (f_\mu, g) = a(f_\mu, U^a g) = a(g, \hat{U}^a f_\mu).$$

Hence for any $a > 0$, we have

(12) $$f_\mu = a\hat{U}^a f_\mu \quad \text{m a.e.}$$

for all $a > 0$. So $\hat{U}^a(f_\mu) < \infty$ m a.e. for all $a > 0$. Since the functions are cofinely continuous in (12), we have

(13) $$f_\mu = a\hat{U}^a f_\mu$$

and $f_\mu \in \hat{S}$. Hence $t \to \hat{P}_t f_\mu$ is right continuous. Thus we obtain for all $t > 0$, $f_\mu = \hat{P}_t f_\mu$. Q.E.D.

COROLLARY 1. If $f \in B^+$ and $f < \infty$ m a.e., then f is an invariant function of X if and only if

(14) $$f(x) = aU^a f(x)$$

for all x and $a > 0$.

PROOF. Necessity is evident. From (14) and Proposition 8 on p. 84 in [4], $f \in S$ and $t \to P_t f$ is right continuous. The sufficiency assertion follows from the uniqueness of the Laplace transform. Q.E.D.

COROLLARY 2. Let μ be a sigma-finite measure not charging cofinely open m-copolar sets. It is an invariant measure of X if and only if for every $a > 0$,

(15) $$\mu = a\mu U^a .$$

PROOF. Necessity is evident. By (10), $\mu \ll m$, and we can take $f_\mu = d\mu/dm \in \hat{S}$ such that $f_\mu = \hat{P}_t f_\mu$ m a.e. for every $t > 0$. Letting $g \in B^+$, we obtain

$$\mu(g) = (f_\mu, g) = (\hat{P}_t f_\mu, g) = (f_\mu, P_t g) = \mu(P_t g) .$$

The sufficiency assertion is proved. Q.E.D.

Let $N^t = [X_s: s \geq t]$ and $N = \bigcap_t N^t$. Let $U_x = \{A \in N: P^x(A \Delta \theta_t A) = 0$ for every $t > 0\}$ (Here, Δ is the symmetric difference).

DEFINITION 4. Let $U = \bigcap_{x \in E} U_x$. Sets in U will be called invariant sets.

It is evident that U is a sigma-field, called the invariant sigma field. If for any $A \in U$, $P^x(A)$ is identically zero or one, then U will be called uniformly degenerate.

THEOREM 3. U is uniformly degenerate if and only if the bounded invariant functions of X are constant.

PROOF. Let U be uniformly degenerate, and let f be a bounded invariant function of X. Then $f = P_t f$, and $f(X_t)$ is a bounded P^x-martingale for every x. Since $f \in S$, $f(X_t)$ is right continuous a.s., hence $\lim_{t\to\infty} f(X_t)$ exists a.s. Since U is uniformly degenerate, there is a constant C such that $C = \lim_{t\to\infty} f(X_t)$, so $f(x) = E^x \lim_{t\to\infty} f(X_t) = C$. Necessity is proved. Now suppose that the bounded invariant functions are constants. Let $A \in U$: then $f(x) = P^x(A)$ is an invariant function of X. Hence there is a constant C_A such that $f(x) = C_A$ and there is a sequence (t_n) with t_n increasing to infinity such that

$$P^x(A \Delta \theta_{t_n} A) = 0 \tag{16}$$

for all $n \geq 0$. If $N_t^\circ = \sigma[X_s: s \leq t]$, we have $A \in \bigvee N_t^\circ \; (= \bigvee N_{t_n}^\circ)$, hence

$$I_A = \lim_{n\to\infty} E^x(I_A \mid N_{t_n}^\circ) \quad \text{a.s. } P^x. \tag{17}$$

On the other hand, from (16),

$$E^x(I_A \mid N_{t_n}^\circ) = E^x(\theta_{t_n} A \mid N_{t_n}^\circ) = E^{X(t_n)}(I_A) = f(X_{t_n}) = C_A \text{ a.s.}$$

Thus we have $C_A = I_A$ a.s. P^x and $C_A = 1$ or 0 follows from the equality above. Q.E.D.

REFERENCES

1. R. K. Getoor, Markov Processes: Ray Processes and Right Processes. Lecture Notes in Mathematics 440 Springer-Verlag (1975).

2. R. K. Getoor and M. J. Sharpe, Naturality, standardness and weak duality for Markov processes. Zeit. fur Wahrscheinlichkeitstheorie verw. Geb. 67 (1984).

3. R. K. Getoor and J. Glover, Riesz decompositions in Markov process theory. Trans. AMS 285 (1984).

4. K. L. Chung, Lectures from Markov Processes to Brownian Motion. Springer-Verlag (1982).

Wu Rong
Department of Mathematics
Nankai University
Tianjin
Peoples' Rep. of China

RIGHT BROWNIAN MOTION AND REPRESENTATION OF INITIAL PROBLEM

Z. Zhao

Let $\{X_t^+ : t > 0\}$ be the right Brownian motion on $[0, \infty)$ determined by the transition density: for $x, y \in [0, \infty)$.

$$p^+(t; x, y) = \begin{cases} \frac{y}{x\sqrt{2\pi t}}[\exp(-|x-y|^2/2t) - \exp(-|x+y|^2/2t)], & x > 0 \\ \sqrt{\frac{2}{\pi}}\frac{y^2}{t^{3/2}}\exp(-y^2/2t), & x = 0 \end{cases} \tag{1}$$

This is a Markov process having the tendency moving to the right direction. 0 can be a starting point, but is never reached, i.e., $\{0\}$ is a polar set.

In this paper, we shall use the right Brownian motion to represent the solution of the following initial problem for the 1-dim. Schrödinger equation:

$$\begin{cases} u'' + 2qu = 0 \\ u(0) = 0 \text{ and } u'(0) = 1, \end{cases} \tag{2}$$

where q is a given Borel function on $[0, \infty)$ with $\int_0^a x|q(x)|dx < \infty$ for any $0 < a < \infty$. This condition on q is a little more general than 1-dim. Kato class.

For $0 < b \leq \infty$, we assume that $((0, b), q)$ is locally gaugeable, namely for any $0 < \ell < b$, $E^x[\exp\int_0^{\tau_{(0,\ell)}} q(X_t)dt] < \infty$ for $0 < x < \ell$, where $\{X_t\}$ is the standard 1-dim. Brownian motion and $\tau_{(0,\ell)}$ is the first exit time from $(0, \ell)$. This condition is equivalent to that

$$\sup[\text{Spec}(u'' + 2qu|_{(0,b)})] \leq 0\,.$$

THEOREM.

$$u(x) \equiv \frac{x}{E^0[\exp \int_0^{T_x} q(X_t^+)dt]}, \qquad x \in [0,b)$$

is the solution of the initial problem (2), where T_x is the hitting time on $\{x\}$.

Remark. $u(x)$ only depends on the "data" i.e., the values of q on $(0,x)$. This property matches the feature of the initial problem. Practically, if x is regarded as the real time, then the formula gives a "on-line" procedure for a computer simulation.

Proof. Obviously, $u(0) = 0$. Since the Green function of $\{X_t^+\}$ $G^+(z,y) \leq \frac{2y(z\wedge y)}{z} \leq 2y$, we have

$$\sup_{z\in(0,x)} \left[E^z \int_0^{T_x} |q(X_t^+)|dt\right] \leq 2\int_0^x y|q(y)|dy \downarrow 0 \qquad \text{as } x \downarrow 0\,.$$

Then by the Khasminskii lemma (see [1]), we have

$$E^0[\exp \int_0^{T_x} q(X_t^+)dt] \longrightarrow 1 \qquad \text{as } x \downarrow 0\,.$$

Hence

$$u'(0) = \lim_{x\downarrow 0} \frac{u(x)}{x} = \lim_{x\downarrow 0} \frac{1}{E^0[\exp \int_0^{T_x} q(X_t^+)dt]} = 1\,.$$

Thus u satisfies the initial conditions. We now verify that u is a solution to the equation in (o,b). Since this is a local problem, we need only to check for each $0 < \ell < b$, u is a solution in $(0,\ell)$.

Since $((o,\ell),q)$ is gaugeable, it is known (see [1]) that

$$v(x) \equiv E^x[T_\ell < T_0, \quad \exp \int_0^{T(0,\ell)} q(X_t)\,dt]$$

is a solution in $(0,\ell)$ with $v(0) = 0$ and $v(\ell) = 1$. Hence we need only prove that

$$u(x) = u(\ell)v(x)\,. \tag{3}$$

To prove (3) we need a fact that $\{X_t^+\}$ and $\{X_t | T_\ell < T_0\}$ have the same distributions until T_ℓ, i.e., we shall prove for any bounded Borel function f, $t > 0$ and $o < x < \ell$, we have

$$E^x[t < T_\ell, f(X_t^+)] = E^x[t < T_\ell, f(X_t) | T_\ell < T_0]\,. \tag{4}$$

Since $(t < T_\ell)$ is F_t-measurable, by definition (1), we have

$$\text{The left side of (4)} = \frac{1}{x} E^x[t < T_0,\ t < T_\ell,\ X_t f(X_t)]\,.$$

On the other hand,

$$\text{The right side of (4)} = \frac{E^x[t < T_\ell < T_0,\ f(X_t)]}{P^x(T_\ell < T_0)}$$

$$= \frac{\ell E^x[t < T_0 \wedge T_\ell,\ f(X_t) P^{X_t}(T_\ell < T_0)]}{x}$$

$$= \frac{\ell E^x[t < T_0 \wedge T_\ell, \frac{X_t}{\ell} f(X_t)]}{x} = \frac{1}{x} E^x[t < T_0 \wedge T_\ell,\ X_t\, f(X_t)]\,.$$

Proving (4). By (4) we have

$$\begin{aligned} v(x) &= P^x(T_\ell < T_0) E^x[\exp \int_0^{T(0,\ell)} q(X_t)dt \mid T_\ell < T_0] \\ &= \frac{x}{\ell} E^x[\exp \int_0^{T_\ell} q(X_t^+)dt]\,. \end{aligned} \tag{5}$$

By the strong Markov property of $\{X_t^+\}$, we have for $0 < x < \ell$,

$$E^0[\exp \int_0^{T_\ell} q(X_t^+)dt] = E^0[\exp \int_0^{T_x} q(X_t^+)dt] E^x[\exp \int_0^{T_\ell} q(X_t^+)dt]\,. \tag{6}$$

Thus (3) follows from (5) and (6). ∎

Reference

[1] K.-L. Chung and Z. Zhao, forthcoming monograph.

Z. Zhao

Department of Mathematics

University of Missouri—Columbia

Columbia, MO 65211

Progress in Probability

Edited by:

Professor Thomas M. Liggett
Department of Mathematics
University of California
Los Angeles, CA 90024-1555

Professor Charles Newman
Department of Mathematics
University of Arizona
Tucson, AZ 85721

Professor Loren Pitt
Department of Mathematics
University of Virginia
Charlottesville, VA 22903-3199

Progress in Probability includes all aspects of probability theory and stochastic processes, as well as their connections with and applications to other areas such as mathematical statistics and statistical physics. Each volume presents an in-depth look at a specific subject, concentrating on recent research developments. Some volumes are research monographs, while others will consist of collections of papers on a particular topic.

Proposals should be sent directly to the series editors or to Birkhäuser Boston, 675 Massachusetts Avenue, Suite 601, Cambridge, MA 02139.

1 ÇINLAR/CHUNG/GETOOR. Seminar on Stochastic Processes, 1981
2 KESTEN. Percolation Theory for Mathematicians
3 ASMUSSEN/HERING. Branching Processes
4 CHUNG/WILLIAMS. Introduction to Stochastic Integration
5 ÇINLAR/CHUNG/GETOOR. Seminar on Stochastic Processes, 1982
6 BLOOMFIELD/STEIGER. Least Absolute Deviation
7 ÇINLAR/CHUNG/GETOOR. Seminar on Stochastic Processes, 1983
8 BOUGEROL/LACROIX. Products of Random Matrices with Application to Schrödinger Operator
9 ÇINLAR/CHUNG/GETOOR. Seminar on Stochastic Processes, 1984
10 KIFER. Ergodic Theory of Random Transformations
11 EBERLEIN/TAQQU. Dependence in Probability and Statistics
12 ÇINLAR/CHUNG/GETOOR. Seminar on Stochastic Processes, 1985
13 ÇINLAR/CHUNG/GETOOR/GLOVER. Seminar on Stochastic Processes, 1986
14 DEVROYE. A Course in Density Estimation
15 ÇINLAR/CHUNG/GETOOR/GLOVER. Seminar on Stochastic Processes, 1987
16 KIFER. Random Perturbations of Dynamical Systems
17 ÇINLAR/CHUNG/GETOOR/GLOVER. Seminar on Stochastic Processes, 1988